The Institute of Biology's
Studies in Biology no. 127

Invertebrate Respiration

Rufus M. G. Wells

M.Sc., Ph.D.

Lecturer in Zoology,
University of Auckland,
New Zealand

49,202

Edward Arnold

First published 1980
by Edward Arnold (Publishers) Limited
41 Bedford Square, London WC1 3DQ

British Library Cataloguing in Publication Data

Wells, Rufus M G
  Invertebrate respiration. – (Institution of
  Biology. Studies in biology; no. 127
  ISSN 0537–9024).
  1. Invertebrates – Physiology   2. Respiration
  592'.01'2   QL364

  ISBN 0–7131–2806–2

To R. P. Dales

Photo Typeset by The Macmillan Co. of India Ltd
Bangalore
and
printed and bound in Great Britain at
The Camelot Press Ltd, Southampton

General Preface to the Series

Because it is no longer possible for one textbook to cover the whole field of biology while remaining sufficiently up to date, the Institute of Biology has sponsored this series so that teachers and students can learn about significant developments. The enthusiastic acceptance of 'Studies in Biology' shows that the books are providing authoritative views of biological topics.

The features of the series include the attention given to methods, the selected list of books for further reading and, wherever possible, suggestions for practical work.

Readers' comments will be welcomed by the Education Officer of the Institute.

1980

Institute of Biology
41 Queen's Gate
London SW7 5HU

Preface

The invertebrates present us with a diversity of adaptive strategies to the problem of balancing an animal's metabolic requirements with its accessibility to oxygen. This book represents an attempt to use these strategies in illustrating fundamental principles of respiratory physiology without an overburden of purely physical or chemical discourse. An encyclopaedic account of respiratory adaptations was resisted and the examples were chosen to reflect recent research trends. I have considered the invertebrates not as 'lower' animals trying inefficiently to copy the vertebrates, but as organisms with control and regulatory capabilities which have to cope with large environmental exigencies.

Auckland, 1980   R.M.G.W

Contents

1 Principles of Gaseous Exchange

From the smallest unicell to the most complex metazoan, an organism presents all or some part of its body as a surface for gaseous exchange. This is necessary because the organism must be furnished with a steady supply of oxygen from the environment and liberate carbon dioxide into it in order to meet its energetic requirements. Thus, when air or water containing a supply of oxygen is constantly passing an organism of lower oxygen and greater carbon dioxide content, the potential for regular exchange is established. Two fundamental questions must be asked and we shall attempt to find answers to them. Firstly, is there a universal mechanism for the transfer of respiratory gases across animal exchange surfaces? If there is, then does the nature of the medium, air or water, dictate constraints upon respiratory strategies? In this chapter we consider the means by which the respiratory gases are transferred between the environment and the organism, and between the organism and its cells. The general principles of gaseous exchange apply to all animals, not merely to those functioning within a narrow range of physical variables. We must expect that a transfer system will meet the oxygen requirements of an animal living in a particular habitat and ensure an adequate rate of carbon dioxide removal.

1.1 Definitions and explanations

1.1.1 Composition of the earth's atmosphere

With the exception of water vapour, the atmosphere remains constant in its composition (see Table 1). These values may be slightly modified in the near-environment due to respiration, photosynthesis and pollution.

1.1.2 Quantity of a gas

From his studies of gases, Avogadro came to the conclusion that equal volumes of different gases at the same temperature and pressure contained equal numbers of molecules. It follows that a *mol* (or mole) of one gas occupies the same volume as a *mol* of any other gas, provided that it is at the same temperature and pressure. For this purpose, gas volumes may be recorded at a standard temperature and pressure (s.t.p.) defined as 273 K and standard atmosphere pressure (101 325 Pa). Thus, the amount, M, of a dry gas may be expressed in mol or volume (s.t.p.). Both conventions are used because the molar volume of an ideal gas is

Table 1 Composition of dry air and the partial pressure exerted by each component at s.t.p. The nitrogen and noble gas fractions may be considered physiologically inert insofar as they do not participate in metabolic reactions. The carbon dioxide fraction may be slightly higher in cities.

	%	p (kPa)
Oxygen	20.95	21.23
Carbon dioxide	0.03	0.03
Nitrogen	78.09	79.13
Argon	0.93	0.94
	100.00	101.33

22.4 dm^3; oxygen is an ideal gas (22.393 dm^3) and carbon dioxide is nearly so (22.262 dm^3). Consequently, the comparison of a quantity of carbon dioxide with a quantity of oxygen expressed in mol is precise, but it is slightly in error if the quantities are expressed in volumes s.t.p.

1.1.3   Concentration of a gas

The concentration C, of a gas is defined for all media (gas, water, body fluids) by the ratio:

$$C = \frac{M}{V}$$

where M is the quantity of gas contained in a unit volume V. Its dimension therefore is (quantity of gas) × (volume)$^{-1}$. In a mixture of gases, each component exerts a partial pressure independently of other gases present. Oxygen and carbon dioxide dissolved in water may be expressed either as concentration (mmol dm^{-3} or cm^3 dm^{-3}) or as a partial pressure of the gas phase in equilibrium with the water (kPa). The term gas *tension* is sometimes used in preference to partial pressure. Both terms are interchangeable and share the common symbol p (see MORRIS, 1974). Some useful conversions are:

Oxygen		Carbon dioxide	
1 dm^3	= 1429 mg	1 dm^3	= 1964 mg
1 mg	= 0.031 mmol	1 mg	= 0.023 mmol
1 mmol	= 22.4 cm^3	1 mmol	= 22.4 cm^3

It will now be apparent that if the partial pressure of the gas above a liquid is reduced, then the gas tension in the liquid will decrease proportionately until an equilibrium is established.

An example, by way of explanation, will serve to illustrate how a body of water in contact with a normal atmosphere, comes to equilibrium. At one atmosphere (101.33 kPa) and 20°C, water vapour contributes a pressure of 2.33 kPa. Using the fraction of oxygen in air, the pO_2 in the water will therefore be $(101.33 - 2.33) \times 0.2095 = 20.74$ kPa, a value slightly less than the 21.23 kPa in air given in Table 1 because the water vapour pressure has been subtracted. pN_2 and pCO_2 may be calculated similarly using the appropriate air-fraction values. Invertebrates are rarely found in environmental pO_2 higher than 21 kPa but many species are found living at low pO_2, some even approaching zero levels.

1.1.4   Solubility and capacitance

Capacitance is a general term including not only physical *solubility* but also chemical binding of gas species in a liquid. This distinction is physiologically important because, as we shall see later, oxygen may be transported in combination with a respiratory pigment. The capacitance, β, of a gas is defined as the change in its concentration (ΔC) per unit pressure change (ΔP) and is quantified by Henry's Law:

$$\beta = \frac{\Delta C}{\Delta P}$$

and the dimension for β is (quantity of substance) $\times$ (volume)$^{-1}$ $\times$ (pressure)$^{-1}$. Values for oxygen and carbon dioxide capacitances in pure water and sea water, and at various temperatures are given in Table 2. These figures show four important features. Firstly, βCO_2 is much higher than βO_2 for water, a property which has important physiological consequences. Although the solubility of carbon dioxide is approximately 30 times greater than that for oxygen, the amount in the atmosphere is very small. The capacitance of carbon dioxide in water is further augmented by the fact that the gas may be present in carbonate and bicarbonate forms. Fluctuations in dissolved oxygen and carbon

Table 2   Capacitance values (β) of oxygen and carbon dioxide in air and water at various temperatures. The dimension of β is μmol dm^{-3} kPa^{-1}. In air, β is the same for oxygen and carbon dioxide, and for water, β is equivalent to solubility (α).

	Distilled water			Sea water		Air
$T °C$	βO_2	βCO_2	$\beta O_2/\beta CO_2$	βO_2	βCO_2	β
0	21.5	759.4	0.028	17.1	637.5	440.6
10	16.7	529.3	0.032	13.7	450.0	425.0
20	13.6	389.2	0.035	11.5	335.6	410.5
30	11.5	294.8	0.039	9.9	258.5	396.9

dioxide in aquatic environments may be quite marked as a result of the respiratory and photosynthetic activities of organisms. Secondly, an increase in temperature decreases β for both gases, which is the reverse of the effect of temperature on the solubility of solids. The decrease in β is more marked for carbon dioxide; one has only to recall the effect of opening a bottle of warm carbonated soft-drink! Thirdly, salt depresses the capacitance and, lastly, in the gaseous phase β is the same for any ideal gas.

Our definitions of concentration and capacitance are valid for the gaseous phase as well as for water and body fluids but they call for some explanation. By way of example, when a volume of air enters the spiracles of an insect at a given temperature and pressure, the fraction of oxygen in the air will be 0.2095. But the concentration of gas in the tracheal endings may change if the pressure and temperature at these sites is different from the ambient conditions.

1.2   Fick's Law – a model for diffusion

Now that we have defined and explained the relationships between concentration, pressure, and capacitance, we can apply these concepts to develop a model system which attempts to explain the movements of the respiratory gases. The merit of such a model will rest on its ability to explain and predict the transfer of oxygen and carbon dioxide in real animals.

Consider the model in Fig. 1–1 in which a gas permeable membrane of

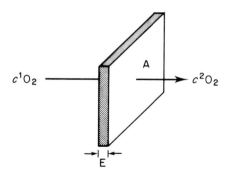

Fig. 1–1   Model representing a system for the description of oxygen diffusion across a gas permeable membrane of surface area A and thickness E. c^1O_2 is the concentration of oxygen and is higher than c^2O_2, and the capacitance $\beta^1O_2 = \beta^2O_2$. Oxygen diffuses in the direction indicated by the arrow until the equilibrium $\Delta cO_2 = 0$ is reached.

cross-sectional area A and thickness E separates oxygen at high concentration c^1O_2 from a lower concentration c^2O_2. Assuming that the gas on both sides of the membrane are well-mixed, then oxygen will diffuse through the membrane from the higher to the lower concentration until equilibrium is established. This movement of gas, called gas *flux*, is described quantitatively by Fick's Law of diffusion which states that during a given period of time Δt, the amount of oxygen MO_2 diffusing through the membrane is directly proportional to the concentration difference $(\Delta cO_2 = c^1O_2 - c^2O_2)$ and the area A of the membrane, but inversely proportional to the membrane thickness E. The Fick equation may be symbolized:

$$MO_2/\Delta t \propto \Delta cO_2 \cdot \frac{A}{E} \tag{1}$$

$$= DO_2 \cdot \Delta cO_2 \cdot \frac{A}{E} \tag{2}$$

where the constant DO_2 is the *diffusion coefficient* (sometimes called *diffusivity*) of oxygen. The dimension of D is (area) × (time)$^{-1}$, usually $cm^2\ sec^{-1}$.

It is important to remember that the value of D depends on the geometry of the membrane and the species of the gas; the properties of a membrane which could enhance diffusion are a large surface area (A) and a small thickness (E). The Fick equation is valid for the conditions stated and when there is the same medium in both compartments. In this example β^1O_2 and β^2O_2 are identical and the *driving force* for diffusion is the concentration difference ΔcO_2. Moreover, Fick's Law remains valid if the membrane is replaced with a small hole. Thus the equation would adequately describe the movements of oxygen and carbon dioxide in the tracheal system of an insect or through the narrow lung opening or 'pneumostome' of a snail.

In living animals however, we rarely know the thickness of the membranes through which gases are exchanged. Sections through preserved specimens are often distorted or contracted, and may not accurately reflect living dimensions, nor is it known whether E remains constant in life. In such cases, it may be useful to consider the constant D/E, the *permeability coefficient*.

One important assumption of Fick's Law does not apply to physiological systems. That is, the media on either side of the membrane are rarely similar and thus usually have different capacitances. What happens when gas diffuses between the following: air/chitin, water/epidermis, blood/water? We must now reconsider the Fick model for a situation where different media are separated by a gas permeable membrane.

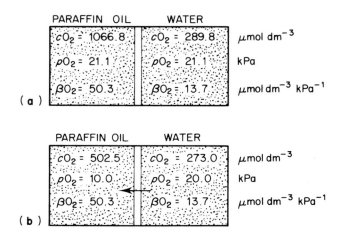

(a)

(b)

Fig. 1–2 (a) An oil compartment and a water compartment with differing concentrations and capacitances of oxygen separated by a gas permeable membrane. Under these conditions there is no net flux of oxygen because there is no pressure difference. (b) Oil and water compartments as in (a) but with a gas pressure difference across the membrane. Under these conditions, oxygen diffuses against the concentration gradient, but with the pressure gradient until equilibrium when the pressure is the same on both sides of the membrane.

In Fig. 1–2 a divided chamber (a) contains oil on one side and water on the other side of a membrane. Both liquids are initially equilibrated with air at atmospheric pressure and the pO_2 is therefore 21.1 kPa on each side. However, the solubility of oxygen is much greater in oil, thus the concentration, as predicted by Henry's Law, is 3.7 times greater than that in the water compartment. In spite of the concentration differences, the two solutions are in equilibrium, with the same pO_2. A different situation is shown in chamber (b) where in addition to a concentration difference there is a partial pressure difference. In this case, the net flux of oxygen is *against* the concentration gradient, but *with* the pressure gradient. (Recalling Henry's Law, this situation is permitted because the capacitance in water is much lower than in oil.) Diffusion proceeds against the concentration gradient until the pO_2 on both sides of the membrane is the same. Thus the driving force for diffusion between media of different capacitances is Δp and it operates against a progressively increasing Δc. We may now introduce Δp into equation 2 by substituting $\Delta p \cdot \beta$ for Δc.

$$MO_2/\Delta t = DO_2 \cdot \Delta pO_2 \cdot \beta O_2 \cdot \frac{A}{E} \qquad (3)$$

$$\text{or} \quad DO_2 \cdot \beta O_2 = \frac{MO_2}{\Delta t} \cdot \frac{1}{\Delta pO_2} \cdot \frac{E}{A} \tag{4}$$

where the product $D \cdot \beta$ defines Krogh's constant of diffusion, K, and may be used where capacitance is unknown. If neither the geometry of the membrane nor the capacitance is known, then all the constants may be combined to give the *diffusive conductance* G where,

$$GO_2 = DO_2 \cdot \beta O_2 \cdot \frac{A}{E} = \frac{MO_2}{\Delta t \cdot \Delta pO_2} \tag{5}$$

1.3 Values of coefficients

Some examples of the coefficients in different media are given in Table 3 and serve to quantify and summarize the statements made in sections 1.2 and 1.3.

Table 3 Values of diffusion coefficient, $D\,(cm^2\,sec^{-1})$; capacitance coefficient, $\beta\,(\mu mol\,dm^{-3}\,kPa^{-1})$ and Krogh's constant $K\,(\mu mol\,cm^{-1}\,sec^{-1}\,kPa^{-1})$ (i.e. the product $D{\cdot}\beta$) for oxygen and carbon dioxide in various media at 20° C.

	DO_2	βO_2	KO_2	DCO_2	βCO_2	KCO_2
Distilled water	0.000 025	13.6	0.34	0.000 018	389.2	7.00
Muscle	0.000 0075	14.0	0.11	0.000 011	345.0	3.80
Chitin			0.01			
Air	0.25	410.5		0.20	410.5	

If we compare oxygen with carbon dioxide we find that the latter has a great advantage. The capacitance of carbon dioxide in water is approximately 28 times greater than that of oxygen, hence CO_2 can be rapidly eliminated in solution. Krogh measured the constant K for various media and found it to be lower in tissues than in water at the same temperature because the diffusion coefficient, D, is reduced in the presence of cellular structures and solutes.

We also see that in a given medium the diffusion coefficient D is similar for both gases, being slightly smaller for carbon dioxide as predicted by Graham's Law because of its higher molecular weight. When air and water are compared, we see that D is much higher in air than in water because the kinetic energy of molecules in the gaseous phase is so much greater. There remains some uncertainty about what values should be used, for instance between blood and cells, within cells, and between groups of cells.

1.4   Facilitated diffusion

In the presence of certain 'carrier' molecules, respiratory gases diffuse at a faster rate than would be expected from the nature of the gases and of the diffusing media. In other words, the rate of gas transport is greater than predicted by Δc or Δp and the effective value of Krogh's constant is modified. Haemoglobin, myoglobin, and other respiratory pigments act as carriers by virtue of their reversible combination with oxygen.

An early explanation for facilitated diffusion was advanced by Scholander. He postulated a 'bucket-brigade' mechanism whereby oxygen was passed from one haemoglobin molecule to another along a decreasing pO_2 gradient. However, recent work has shown that Scholander's scheme is inconsistent with experimental observations and facilitated diffusion may be fully explained by the simultaneous diffusion of oxygen in physical solution *and* in combination with haemoglobin. Haemoglobin aids net oxygen transport only if there is an oxyhaemoglobin gradient and a pO_2 gradient. The pO_2 gradient is dependent only on oxygen in physical solution and is independent of the presence of 'carrier' molecules like haemoglobin. If the haemoglobin is chemically modified to impair its reversible combination with oxygen, then diffusion is not facilitated.

It may seem odd that a large, slowly diffusing haemoglobin molecule (Hb) enhances the uptake of a rapidly diffusing oxygen molecule. It will be recalled that diffusion flux is a function of both $\Delta cHbO_2$ and βHbO_2 thus although $cHbO_2$ may be 100 times greater than the concentration of molecular oxygen, the fluxes may be of the same order of magnitude. Similarly, the increased movement of carbon dioxide in the presence of carbonic anhydrase may be explained by the diffusion of bicarbonate ions in addition to carbon dioxide *per se*. The facilitation of bicarbonate transport by the enzyme is analogous to the haemoglobin mechanism through its catalytic action in the reaction:

$$CO_2 + H_2O \rightleftharpoons H^+ + HCO_3^-$$

The physiological importance of facilitated diffusion is only speculative because the physical conditions in living animals are unknown and experiments have not been carried out under these conditions. Undoubtedly facilitated diffusion is an important process in oxygen and carbon dioxide diffusion in invertebrates because the distribution of carbonic anhydrase and respiratory pigments for oxygen transport are widespread among the various phyla. The process is likely to be important in accelerating the diffusion of oxygen towards mitochondria in muscle containing myoglobin where the demand for oxygen is periodically very high. It may also be important in accelerating oxygen uptake and carbon dioxide loss in erythrocytes (which are rich in haemoglobin and carbonic anhydrase) at respiratory surfaces as well as in the reverse reactions at the level of actively respiring tissues.

1.5   Importance of gaseous exchange by diffusion

In very small invertebrates the diffusion distances are small and the surface area to volume ratio is large. For this reason, diffusion alone is sufficient for the transfer of gases. An increase in size necessitates convective mechanisms such as irrigation, ventilation or circulation, to augment diffusion.

The insects, are among the most complex metazoans, however gas transport is effected by diffusion alone throughout the hollow, tubular tracheal system. In this case, the advantage of breathing air is clear because the value of D in air is 10 000 times greater than in water. But because diffusion distances are long, and the demands for oxygen in insects are high, diffusion itself imposes limits to growth. In the unlikely event that scientists are ever able to breed gigantic insects, then the restrictions of oxygen demand would permit a metabolic rate for activity at a mere snail's pace!

Isopods (slaters) are a group of arthropods which are restricted to damp terrestrial or aquatic habitats and are usually just a few mm in length. The author recently had the opportunity to study the respiratory physiology of the giant isopod *Glyptonotus*, found under the sea ice of Antarctica. The huge size of this isopod (*ca.* 120 mm) is presumably permitted by the high solubility of oxygen at the low $(-1.9°C)$ temperature at which this animal lives, coupled with a sluggish level of activity not at all characteristic of temperate garden slaters.

While the evolution of terrestrial invertebrates has been spectacular, the largest number and diversity of invertebrates, especially less highly organized ones, are aquatic. It is important to remember that for air-breathers, oxygen and carbon dioxide are approximately equivalent but in water-breathing invertebrates, carbon dioxide, having a high capacitance, is always at a low partial pressure, even when the extraction of oxygen from the water is high.

Some examples illustrate the importance of gas exchanges by diffusion for various invertebrate tissues and different media and are summarized below.

(i) Direct diffusion between water and cells; in Protozoa, Porifera, Coelenterata, many Platyhelminthes, Nematoda, Rotifera, a few Annelida and Bryozoa where oxygen is obtained directly from the surrounding water.

(ii) Direct diffusion between air and cells; the insect tracheal system.

(iii) Diffusion between water and blood; through gills of many aquatic polychaetes and crustaceans.

(iv) Diffusion between air and blood; through wet skin of soft-bodied terrestrial invertebrates, 'book' lungs in spiders, diffusion lungs in pulmonate snails.
(v) Diffusion between blood and cells in tissues.
(vi) Diffusion between intracellular structures; the ultimate fate of oxygen is in the electron transport chain in mitochondria. Effect of cytoplasmic streaming on exchanges entirely unknown.

1.6 Conclusion

In the early part of the century there were physiologists such as C. Bohr and J. S. Haldane who believed that some tissues could actively secrete oxygen and carbon dioxide against a pressure gradient. All the biological evidence available to us now however, suggests that respiratory gases always move across an interface by passive diffusion. Therefore, we can answer the first question posed at the beginning of the chapter and state that diffusion is the universal mechanism which accounts for the movement of gases across an exchange surface. The driving force for the diffusion of a gas is a gradient from high to low partial pressure of that gas. The driving force tends to run 'downhill' until there is no difference in partial pressure and the system is in equilibrium.

The advantage of terrestrial living is that the rate of diffusion is much higher in air than in water and thus steep gradients are more easily maintained, often by bulk gas transport brought about by ventilation of respiratory surfaces and by circulation of body fluids. This only partially answers the second question we asked earlier because further analysis is complicated by the presence of simultaneous oxidations and reductions in the energy-yielding reactions of the cells. Additional complications arise when the exchange surfaces themselves are moving since two pressure gradients have to be considered.

We know very little about the final diffusion pathways for any invertebrate, nor indeed what values we should employ for diffusion coefficients. Little progress has been made since August Krogh's estimations were made in 1919.

2 Surfaces for Gaseous Exchange

An animal presents all or some part of its body surface to the environment for gaseous exchange. The importance of area and thickness (or distance) of exchange surfaces for diffusion has already been discussed using models to describe Fick's equation. In real animals, tissue diffusion distances of more than 1 mm are exceptional because the rate of diffusion is too slow to meet the oxygen requirements of respiring cells. Many small invertebrates, up to a few mg in weight, exchange oxygen and carbon dioxide with the surrounding medium through their entire body surfaces. When diffusion distances are less than 1 mm, surface area does not limit body size by respiratory considerations, because all parts of the body can exchange gases with the ambient medium. There are many unicellular organisms of spherical shape whose area to volume ratio has the lowest theoretical value, but for larger animals with more organized structure, any deviation from the spherical shape improves the ratio for gaseous exchange. Thus, the lengthening and flattening of nematodes, flatworms and other invertebrates may be viewed as respiratory adaptations.

There are two widespread strategies of surface area specialization, each of which is dictated by the medium. In water, evaginations of the body surface constitute *gills* and are principally adaptations for aquatic exchange. In air, invaginated surfaces or *lungs*, are internally protected and link respiring tissues with the atmosphere. In either case, the development of specialized areas for exchange goes hand-in-hand with the development of a convective system for internal transport of gases to and from the exchange surfaces.

A question which might be raised is why, in air, is an invaginated surface more effective than an evaginated one? It is often asserted that in air there is a continual conflict between the need for oxygen uptake and the need for water conservation and that internal surfaces reduce excessive water loss. However, it ought to be pointed out that a surface reducing diffusive losses of water vapour must also reduce oxygen gains. In the absence of comparative data of water and oxygen utilization efficiencies it is difficult to support the assertion. It may be that lungs simply offer protection from mechanical damage in a terrestrial environment.

2.1 Integuments

Apart from a few exceptions, like the exoskeleton of arthropods, the molluscan shell, and calcareous ossicles of some echinoderms, the ordinary body surface contributes to some degree in gaseous exchange. Many aquatic organisms, including Protozoa, Porifera, Coelenterata, Platyhelminthes, Bryozoa, Nematoda, Rotifera and some Annelida, have no specialized respiratory organs and the demands for oxygen are met by diffusion across the general body surface. Some of these invertebrates have channels through which water passes. For example, in sponges there is a large area of naked cell surface available for exchange along the inhalent canals and at the cellular surfaces of the choanocytes.

The integuments in Oligochaeta, Hirudinea, and some Polychaeta are highly vascular and provide the principal route for exchange. Integumentary respiration is also important in the Sipuncula, Echiurida, and some aquatic insects. In the Gastropoda and Polychaeta with specialized organs for respiration, the integument may provide up to half an animal's oxygen requirements.

2.2 Gills

Gills, sometimes called *branchiae*, are a common solution to the problem of gaseous exchange in aquatic invertebrates. Oxygen-depleted blood or coelomic fluid flowing into these organs encounters oxygen at a higher partial pressure diffusing across a thin epithelium. Carbon dioxide diffuses in the opposite direction because its partial pressure is lower in the ambient medium.

A potential problem arises when oxygen has been extracted from the layer of water immediately in contact with gill surfaces. Under these circumstances, the effective surface area of the gill tends to be reduced to a *boundary layer*, and further extraction is limited by the diffusion of oxygen from the medium at some distance from the exchanger. For a gill to function efficiently, some form of convection is necessary to reduce the thickness of the boundary layer. In many cases, gross water currents or turbulence serve this purpose, but invertebrates living in slow-moving or stagnant water must stir up the medium for themselves.

Although there is a great variety of gill forms, the convective mechanisms fall into two categories. Firstly, ciliated surfaces, often organized into well-defined tracts, may induce a continuous flow of water over respiratory surfaces. Cilia on the ctenidia of bivalves, gastropods and ascidians, on the lophophore of brachiopods, the branchial crown of some tubicolous polychaetes like *Sabella*, or on the parapodia of errant polychaetes like *Nephtys*, all serve to ventilate respiratory surfaces. Secondly, muscular movements of the whole animal, or its appendages, including the gills themselves, induce water flow. Peristaltic waves in the body muscles of terebellid polychaetes and chironomid (midge) larvae, rhythmically beating gills in aquatic insect larvae and the polychaete *Chaetopterus*, and the pumping action of the scaphognathites in decapod Crustacea are examples of water flow generated by muscular contractions. In the cephalopods, a muscular mantle cavity with inlet and outlet valves forms an efficient ventilatory pump, also serving in locomotion.

Abundant 'tube feet' or *podia*, and *papulae* are extensions of the body surface in echinoid and asteroid Echinodermata. These organs function as extremely simple gills which nevertheless sustain modest gradients of oxygen transfer between the water vascular system and a well-aerated medium.

Among the insectan orders, the Trichoptera, Plecoptera, and Ephemeroptera have aquatic larval stages with paired gills provided with numerous branching trachea which transport oxygen to respiring tissues. Tracheal gills of larval Odonata lie in an enlarged rectum, forming a branchial chamber ventilated by muscular movements of the abdomen. Metamorphosis into air-breathing adults thus imposes a unique constraint in insect respiration in that an aerial transport system has been modified for aquatic exchange during early development.

Active invertebrates like the Crustacea and some Polychaeta have blood with a respiratory pigment entering the gills in afferent vessels where carbon dioxide is discharged and, having become fully charged with oxygen, the blood leaves the gills by efferent vessels to replenish the tissues.

The efficiency of gaseous exchange through gill surfaces between the medium and body fluids is obviously dependent on the flow rates of the two fluids. Aside from the *rate of flow*, the *direction* of water flow in relation to blood flow is important in the efficiency of gaseous exchange. In the case of echinoderm tube feet the efficiency will not be very high because the transfer of gas is diffusion dependent and extraction proceeds so slowly that only a low metabolic rate can be sustained. Where fluids on either side of an exchanger are in motion, the extraction efficiency becomes *perfusion* dependent. Let us consider the exchange systems in Fig. 2–1. In the concurrent exchanger, deoxygenated blood enters the gill and meets water flowing in the same direction on the other side of the epithelium. The blood gains and the water loses oxygen along the length of the exchanger such that the pO_2 of both fluids is averaged when the flows divert. In the second model with crosscurrent exchange, deoxygenated blood having the first contact with water reaches a higher pO_2 than does blood entering the exchanger at some distance from the initial contact. In the crosscurrent exchanger, the oxygen content of efferent blood will have a higher mean pO_2 than in the concurrent

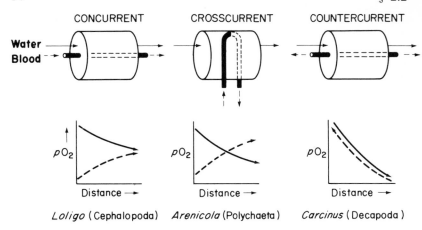

Fig. 2–1 Three 'pipe-within-a-pipe' models to explain gaseous exchange in invertebrate gills. The medium is represented by cylinders of large diameter, and its flow by solid arrows. Smaller pipes and broken arrows represent blood and its direction of flow. The efficiency of the models is compared in the graphs where extraction of oxygen is seen to be greatest in the countercurrent flow because blood leaves the gill at the pO_2 of the incurrent water.

system. The third and most efficient exchanger is the countercurrent system, where deoxygenated blood comes into contact with oxygen in water of progressively higher pO_2 along the length of the transfer surface. The flow rate of blood relative to water is clearly very much greater in the countercurrent system and the result is that efferent blood leaves the exchanger with a much higher pO_2 than that in excurrent water.

A simple experiment demonstrates the importance of gills in respiration. The gills of an invertebrate are ligated or removed and subsequent measurements of oxygen uptake are compared with those from control specimens. In the hemipteran bug *Corixa*, oxygen uptake was reduced by nearly ninety per cent following gill extirpation. Such experiments however, do not take account of possible changes in respiration rates due to trauma.

Observations of decapod Crustacea provide further evidence for the role of gills in aquatic respiration. In wholly aquatic crabs both the gill number and gill surface area are greater than in intertidal crabs and these in turn, are greater than those in fully terrestrial species. Special problems attend the exposure of gill-breathers to air, despite the greater oxygen content and low energetic cost for ventilation in terrestrial environments. The tendency in air is for finely divided gill lamellae to coalesce and considerably reduce the area for gaseous exchange. Furthermore, the problem of osmotic stress might arise through desiccation of the gills, and, in the longer term irreversible damage could occur to the delicate exchange surfaces. In arthropod gills, a chitinous covering provides support for the transition to air-breathing, but imposes an additional barrier to diffusion. The gills of crabs are protected, to some extent, from desiccation and mechanical damage by enclosure within lateral extensions (branchiostegites) of the carapace. Consequently, gill respiration is not a strategy frequently encountered on land.

A few invertebrates, like amphipods and isopods, have gills which serve a respiratory function alone, but because gills are often in direct contact with ambient conditions, they may have other functions. Gills generate feeding currents for bivalves, some polychaetes, *Daphnia* and *Artemia*. They function in ionic regulation in many arthropods, especially in estuarine and freshwater habitats. Ventilation of gills in the mantle chamber of cephalopod molluscs serves also a locomotor function. And in many invertebrates, the gill is a sensory surface conveying information about the environment to the central nervous system.

In summary, the essential features of gill exchangers are a large surface area, very thin integument, and a mechanism for renewing water at the boundary layer. Ultimately, the establishment of body fluid circulation improves the efficiency of oxygen extraction. Precise measurements of gill surface area in invertebrates are difficult to obtain and usually depend on approximations of the surface geometry for calculation. Recent investigations of gill surfaces with a scanning electron microscope have revealed an extensive microsurface composed of numerous microvilli (Fig. 2–2). The significance of this finding for gaseous exchange has not yet been fully assessed although it is likely that it will have to be taken into account when comparing gill surface areas from unrelated species.

2.3 Lungs

Clearly, air is the most favourable medium for uptake of oxygen by an animal because of the higher diffusion coefficient. But the advantage of access to a plentiful supply of oxygen on land may be partly offset by the need to conserve water in a dry environment. Invaginated respiratory surfaces present surfaces across which oxygen and carbon dioxide diffuse and with their internal disposition, may reduce evaporation of water.

2.3.1 Diffusion lungs

Many Arachnida (spiders and scorpions) have expanded internal

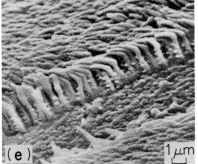

surfaces known as 'book lungs' because of the subdivision of their pulmonary cavities by numerous parallel lamellae. Oxygen diffuses from the atmosphere through narrow tubes, the tracheae, into the lung where it is exchanged for carbon dioxide across the lamellae, with perfusing haemolymph, and then with respiring tissues via the circulation.

In pulmonate snails, the mantle cavity is modified as a lung that communicates with the atmosphere through a single opening, the pneumostome, which may be opened or closed according to respiratory conditions. In terrestrial slugs and snails, carbon dioxide stimulates the opening of the pneumostome. Regular ventilation has never been clearly demonstrated in pulmonates, but locomotory movements of the foot do cause the floor of the mantle to rise and fall with each phase of muscular contraction.

A re-colonization of aquatic environments has occurred in a number of the Pulmonata. The amphibious snail *Lymnaea*, lives in freshwater ponds, often deficient in oxygen, where it takes a volume of air into the lung, using it as a store when foraging under water. Periodically, *Lymnaea* must surface for renewal of its oxygen store. The Ugandan swamp worm *Alma emini* also lives in a habitat of low pO_2. In this case the tip of its flattened tail protrudes above the water and curls around to trap a bubble of air which is brought into contact with a thin well-vascularized area on the dorsal body surface acting as a kind of lung. These adaptations emphasize the improved efficiency by using air compared with slow diffusion in water.

The mantle cavity is an important adaptation in land colonization by the Mollusca. In terrestrial shelled pulmonates, the relative lung surface area is comparable with mammalian lung surfaces; in *Helix pomatia* the lung surface occupies 8.3 cm² per g body weight while for man it is $10.9 \, \text{cm}^2 \, \text{g}^{-1}$. But in the naked slug *Arion*, it is only $0.7 \, \text{cm}^2 \, \text{g}^{-1}$, presumably reflecting a greater ease of integumentary exchange. A unique adaptation in the arboreal New Zealand tree slug,

---

Fig. 2–2　Views of some exchange microsurfaces as seen with a scanning electron microscope. (**a**) Portion of a gill tuft from the marine polychaete *Terebella* showing a large surface area given by branching and folding and (**b**) by a microvillous surface at higher magnification. (**c**) Surface of a tracheal gill from the freshwater mayfly larva *Zephlebia* showing tracheal branching and (**d**) at a higher magnification showing extensive folding of the cuticle. (**e**) Surface portion of the plastron from the fresh water bug *Aphelocheirus* showing the disposition and density of hydrofuge hairs which trap a layer of air next to the integument. (**a**)–(**d**) are original pictures and (**c**) is after Hinton, H. E. (1976) *J. Insect. Physiol.*, **22**, 1529–50.

Athoracophorus, is the reduction of the mantle cavity to a small chamber whose walls are produced into tubules that ramify through adjoining blood spaces in a fashion reminiscent of a tracheal system.

2.4   Tracheae

Undoubtedly the most successful colonizers of the terrestrial environment are the Insecta. Though their integuments are hard and impermeable to gases and a waxy covering reduces water loss, they have air-filled tubes, *tracheae*, developed as ingrowths from slits (spiracles) in the body surface. The tracheal system thus falls within our general definition of 'lungs'. Tracheal systems are also found in onychophorans, chilopods, some spiders, and mites. Exchange between air in the tracheae and the atmosphere is regulated by closure mechanisms of the spiracles. Air in the trachea diffuses through profusely branched tubes to the finest branches of the system, the tracheoles, usually less than 1 μm in diameter and these may actually penetrate cells to bring oxygen into close contact with mitochondria. Gaseous exchange probably occurs throughout the entire length of the tracheal system and in the tracheoles, but because the system is subdivided, the finest branches will collectively present the greatest surface area and be of greatest importance in exchange. While diffusion alone can account for the movement of gas in tracheae, especially in resting insects or ones of small size, a number of larger species ventilate their tubes by thoracic or abdominal movements.

2.5   Physical gills and plastrons

Aquatic insects have evolved from terrestrial forms and, for the most part, have retained the tracheal system for respiration. Adults of a number of species of Hemiptera and Coleoptera spend a great part of their lives in water, surfacing periodically to trap air bubbles against spiracular openings on the body surface for use while the animals are submerged. The diving beetle *Dytiscus* has a large air space under the elytra which serves as a simple store of oxygen which must be renewed when it is exhausted. During a dive, oxygen is consumed and carbon dioxide which is released into the bubble then dissolves in the surrounding medium.

In some Notonectidae (waterboatmen) and Corixidae a greater surface of the air bubble is exposed to water than in *Dytiscus*, and gases are exchanged between the insect and the medium. In this case, the air bubble is held in place against a single pair of spiracles by hydrofuge hairs and functions as a physical gill. The hairs cannot be wetted and so they easily penetrate the surface film of water and put the tracheal system into communication with the atmosphere. When submerged, carbon dioxide dissolves readily in water and, as oxygen is removed, more diffuses in from the water under the difference in partial pressures. This situation might last indefinitely but nitrogen from the bubble also gradually dissolves into surrounding water and the bubble shrinks, its surface area becomes too small for adequate diffusion, and the animal must surface to renew the gill. Leg movements ventilate the bubble during swimming and prevent the formation of an oxygen-deficient boundary layer between the water and the bubble.

The mechanism of physical gills imposes certain restrictions on underwater life because a bubble increases an insect's buoyancy causing it to rise to the surface whether its respiratory needs require it or not. An interesting adaptation in the notonectid genera *Anisops* and *Buenoa* imparts to the physical gill a secondary role of buoyancy control. Short air-filled tracheae penetrate an abdominal mass of tissue rich in haemoglobin, which serves as an oxygen store in addition to a very small air bubble. The bubble overlies spiracles on the ventral thorax from which oxygen is drawn for respiration. From time to time, the bubble is flicked backwards to cover abdominal spiracles to pick up oxygen released from the mass of haemoglobin, and then forwards to the spiracles for respiratory intake. A 20 mg *Anisops* contains approximately 0.37 mg haemoglobin, sufficient to combine with 0.5 mm^3 oxygen at the surface. The bug consumes about 0.074 $mm^3 O_2$ min^{-1} at $20°C$ and dives for periods of about four minutes. During this time underwater, oxygen bound by the haemoglobin is released for respiration and can easily supply the 0.3 mm^3 oxygen (0.074 $mm^3 O_2$ min^{-1} × 4 min) required for respiration. Because the bubble is very small, *Anisops* is held in neutral buoyancy enabling it to remain poised in mid-water and make predatory strikes in a horizontal plane. When oxygen from the haemoglobin is used up, the small bubble of air is rapidly depleted of its oxygen content, buoyancy becomes negative, and the insect surfaces to recharge its gill and saturate the haemoglobin with oxygen for another dive. Other members of the Order lack haemoglobin, have positive buoyancy, and normally attack in a vertical plane.

Provided that a physical gill could withstand compressional forces and the decrease in surface area caused by diffusion of nitrogen from the bubble, then the insect might survive underwater indefinitely. Some less active insects do indeed exploit this strategy and never surface. The naucorid bug *Aphelocheirus* is densely covered with minute hydrofuge hairs (*ca.* 4×10^6 mm^{-2}) through which water cannot force its way at ordinary hydrostatic pressures and the animal is always covered by a thin, silvery layer of gas. This thin, and incompressible gill is called a *plastron* (Fig. 2–2) and functions as a surface for gaseous exchange. The maximum depth to which *Aphelocheirus* can dive is limited by hydrostatic pressure and beyond a pressure of 40 kPa (10 m) the hairs are wetted and the bubble disintegrates.

The surface area of a plastron appears to be extensive but precise measurements of area are difficult to make because its structure varies on different parts of the body. Nevertheless, respiratory measurements have shown that the plastron accounts for about 85 per cent of the oxygen uptake in *Aphelocheirus*, the remaining 15 per cent being provided from integumentary exchange. One disadvantage of plastron respiration is that the insect is limited to well-aerated water because should the ambient oxygen tension fall, then diffusion of oxygen into the plastron may be insufficient to sustain respiration. The distribution of aquatic insects seems to support this statement; those inhabiting temporary, oxygen-deficient waters tend to have physical gills renewable at the surface, while plastrons and tracheal gills are seen in insects from well-oxygenated habitats.

2.6   Water lungs

Few aquatic invertebrates have internal breathing surfaces. In the Holothuria, paired structures lying in the coelom formed from hollow outgrowths of the cloaca function as exchangers. Water lungs, or 'respiratory trees' as they are sometimes termed, are finely branched structures ventilated by alternate contractions of longitudinal and circular muscles in the body wall that close and open a cloacal sphincter thus producing a rhythmical pumping action. Oxygen is transferred from water in the 'trees' across their thin exchange surface and into the coelom where erythrocytes transport oxygen to respiring tissues. In the sea cucumber *Cucumaria*, nearly 60 per cent of the gaseous exchange takes place across the surface of its water lungs. A similar system is found in the gephyrean worm *Urechis caupo*, which ventilates its thin-walled hindgut with sea water, exchanging gas with coelomic erythrocytes. In both *Cucumaria* and *Urechis* the internal respiratory surfaces communicate with the environment through a single opening, the cloaca, and because diffusion is so much slower than in air-filled lungs, ventilation is essential for the replacement of an oxygen deficient boundary layer. In contrast with oxygen, carbon dioxide is removed principally through the soft integument of these invertebrates.

2.7   Models of exchangers

Simple models for diffusion between two media separated by exchangers summarize the essential features of respiratory surfaces in invertebrates. The scheme in Fig. 2–3 invites us to consider some of the

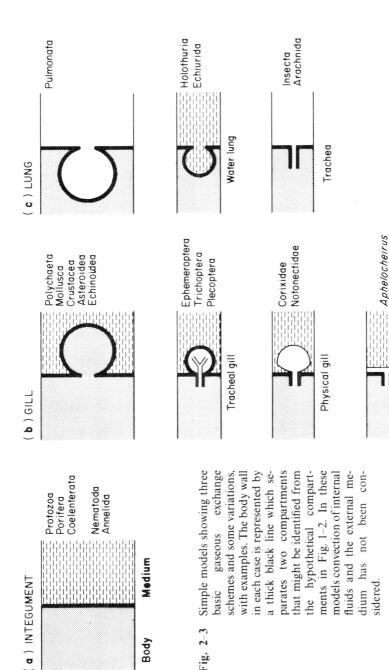

(a) INTEGUMENT

Protozoa
Porifera
Coelenterata

Nematoda
Annelida

Body   Medium

(b) GILL

Polychaeta
Mollusca
Crustacea
Asteroidea
Echinoidea

Ephemeroptera
Trichoptera
Plecoptera

Tracheal gill

Corixidae
Notonectidae

Physical gill

Aphelocheirus

Plastron

(c) LUNG

Pulmonata

Holothuria
Echiurida

Water lung

Insecta
Arachnida

Trachea

Fig. 2–3   Simple models showing three basic gaseous exchange schemes and some variations, with examples. The body wall in each case is represented by a thick black line which separates two compartments that might be identified from the hypothetical compartments in Fig. 1–2. In these models convection of internal fluids and the external medium has not been considered.

features of the hypothetical models discussed in Chapter 1. Real animals may be identified as compartments of different sizes, having diffusion distances, thickness and composition of exchange surfaces, and convective mechanisms. Despite many variations in the kinds of body fluids and exchangers, the number of systems is limited. Within systems there are few taxonomic relationships evident because the systems are adaptive and reflect a particular animal's status in the environment. The exchanger works not in isolation, but in concert with the transport role of the body fluids and in the following chapter we see how this role too, is adaptive.

3 Respiratory Gas Transport by the Blood

3.1 Limits to diffusion

Diffusion alone may account for the movement of oxygen and carbon dioxide between external respiratory surfaces and mitochondria in the smallest organisms provided that the diffusion pathways are less than 1 mm in length. An increase in size, or an increase in activity impose greater demands for oxygen than can be satisfied by simple diffusion, and under these circumstances a convective mechanism must assist the gas fluxes. Some organisms, like sponges and coelenterates, accomplish this by merely stirring up the medium in which they live, but many others, especially those with a highly organized body plan, circulate a portion of their body fluid through a blood vascular system in addition to moving the external medium. The capacitance of a blood, as well as its flow rate, are important factors in determining the amounts of oxygen and carbon dioxide that can be transported. Nearly all the invertebrate phyla have some representatives which possess a respiratory pigment which picks up and delivers oxygen over a given decrement in pO_2. Such pigments may be regarded simply as proteins which effectively increase the capacitance, βO_2, of blood.

3.2 Distribution of respiratory pigments

Each of the four known respiratory pigments may be found in the Invertebrata. They are, in order of the frequency at which they occur: haemoglobin, haemocyanin, haemerythrin, and chlorocruorin. When combined with oxygen they impart respectively a red, blue, pink, or green colour to the body fluids. A tentative classification of the respiratory pigments into an adaptive or physio-ecological pattern is offered in Table 4 with two limitations to be considered. Firstly, we do not have accurate information concerning the levels of oxygen in the micro-environments in which we presume invertebrates to breathe. And secondly, an organism's demand for oxygen as it has been measured in a laboratory respirometer may not accurately portray its natural behaviour.

Nevertheless, we may draw some general conclusions from Table 4:

Table 4 Classification of invertebrate phyla into the three broad groupings according to their scope for activity, access to oxygen, respiratory medium, and the presence of respiratory pigments. Haemoglobin (Hb), haemocyanin (Hcy), chlorocruorin (Chl), haemerythrin (Hr); open circulation (○), closed (●); habitat marine (M), freshwater (FW), or terrestrial (TER).

(a) Narrow scope for activity with easy access to oxygen

Protozoa			M, FW
Porifera			M
Coelenterata			M, FW
Annelida (some)	●		M, FW
Bryozoa	○		M
Brachiopoda (most)	○		M
Mollusca: Bivalvia	○		M, FW
Echinodermata (except Holothuria)	○		M
Mollusca: Amphineura	○	Hcy	M
some Gastropoda	○	Hcy	M, FW, TER
Annelida: *Magelona*	●	Hr	M
Brachiopoda: *Lingula*	○	Hr	M
Annelida: Polychaeta (4 families)	●	Chl	M

(b) Narrow scope for activity with poor access to oxygen

Annelida: Polychaeta	●	Hb	M
Oligochaeta	●	Hb	FW, TER
Echinodermata: Holothuria (some)	○	Hb	M
Mollusca: Gastropoda (some)	○	Hcy	M, FW
Sipuncula	●	Hr	M

(c) Wide scope for activity with easy access to oxygen

Arthropoda: Insecta			TER
Crustacea	○	Hcy	M, FW, TER
Arachnida	○	Hcy	TER
Mollusca: Cephalopoda	●	Hcy	M
(Vertebrata	●	Hb	M, FW, TER)

(d) Wide scope for activity with poor access to oxygen
? None

(i) Invertebrates with low activities and a plentiful supply of oxygen may have no need for a respiratory pigment, especially if they are small in size. These organisms are typically found in marine and freshwater environments.

(ii) Active invertebrates demand good access to oxygen and, with the exception of insects which have no respiratory pigment, they have haemocyanin.

(iii) It is the environmental access to oxygen rather than metabolic demand which correlates with the presence of haemoglobin in invertebrates.

3.3 Structural and functional properties

All of the respiratory pigments are compounds which contain either iron or copper in a protein complex. Because of the coloured nature of these compounds when in combination with oxygen, they are readily characterized by the amount of light they absorb at particular wavelengths in the visible region of the electromagnetic spectrum. Thus when haemoglobin is exposed to air it appears red because it absorbs green light. Spectral differences between oxygenated and deoxygenated pigments are often used to monitor the state of oxygenation in blood.

3.3.1 Haemoglobin and chlorocruorin

The prefix *haem* meant 'blood' in ancient Greek but it is now reserved for the specific chemical entity of iron-porphyrin which in haemoglobin and chlorocruorin is coupled to a protein moiety called 'globin'. Haem is composed of a ring of carbon, hydrogen and nitrogen atoms called porphyrin, with an iron atom at its centre. Porphyrins are a widely distributed group in nature and include the cytochrome enzymes. Haem on its own binds oxygen so tightly that the bond once formed is hard to break. This is because an iron atom can exist in two valency states. Normally, it is the ferrous haem that reacts loosely with oxygen and hence reversibly. When oxidized, haem cannot reversibly combine with oxygen. In haemoglobin, the ferrous haem is protected in a cavity of protein so that oxygen cannot form a tight bond with the iron and the reaction is reversible. Thus we distinguish the reactions of haemoglobin with oxygen as *oxygenation* rather than oxidation, and the reverse process as *deoxygenation* rather than reduction.

The haem is a constant entity in all haemoglobins from man to protozoan, but it is the variation in the amino-acid composition of the globin that gives haemoglobin its dazzling display of functional properties. Like all proteins, haemoglobin and chlorocruorin are made up of specific amino-acid sequences into one or several peptide chains. One and sometimes two peptide chains are combined with a single haem to form a subunit or *monomer*. Haemoglobin may form aggregates from two to more than one hundred subunits with corresponding molecular weights ranging from 34 000 to 3×10^6 daltons.

Haemoglobin may be found as monomers or dimers in the red muscles attached to many molluscan radulas and some invertebrate body wall muscles where it is termed *myoglobin*. Intracellular haemoglobin also occurs in a few protozoa and parasitic trematodes; in the coelomocytes of some polychaetes, clams, and holothurians; in the insect larvae of *Chironomus*; in abdominal sacs of some notonectid insects, and in the parthenogenetic eggs of the crustacean *Daphnia*. In some annelids and molluscs, a neural haemoglobin of uncertain function is found in the sheath of large nerve cells.

Haemoglobins and chlorocruorins found circulating in some invertebrate vascular systems are freely dispersed in the plasma, never in cells, and invariably have a high molecular weight. Extracellular haemoglobins are characteristic of many annelids and other worms, some Entomostraca, and the single gastropod *Planorbis*. These haemoglobins are sometimes called *erythrocruorin* to distinguish them from the intracellular pigments of lower molecular weight. The large molecular size of erythrocruorin may be necessary to reduce the colloidal osmotic pressure of the blood, or to prevent loss through the organs of excretion. Erythrocruorins are sufficiently large that the molecules may be clearly seen when viewed through an electron microscope. An electron micrograph of earthworm erythrocruorin is shown in Fig. 3–1 and the molecules have a two-tiered hexagonal structure with each tier being composed of six units.

Chlorocruorin occurs in just four families of marine polychaetes: Sabellidae, Flabelligeridae, Ampharetidae, Serpulidae; and may be considered a variant of erythrocruorin. Indeed, one genus *Serpula*, has both erythrocruorin and chlorocruorin dissolved in its blood. Chlorocruorin resembles erythrocruorin in having a haem but differs from it by the substitution of a formyl for a vinyl radical on one pyrrole ring of its haem. In addition, it is 'dichroic', that is, it appears green in dilute solution and red when concentrated.

3.3.2   Haemocyanin

The copper protein haemocyanin is present in two phyla only: Mollusca and Arthropoda. In molluscs, it has been found in the Amphineura, Cephalopoda, and many Gastropoda, but it is absent in the Bivalvia. Among the Amphineura and Gastropoda (but not Cephalopoda), there are species that have haemocyanin in the blood and

---

Fig. 3–1   Electron micrographs of (**a**) deoxygenated haemocyanin molecules from the snail *Helix*, and (**b**) when oxygenated. (**c**) haemoglobin molecules from the earthworm *Lumbricus*. Notice that deoxyhaemocyanin is polymerized but, on reaction with oxygen, the molecules dissociate into subunits. The quaternary structure of the haemoglobin is stable and the two-tiered hexagonal symmetry of the molecules can be discerned.

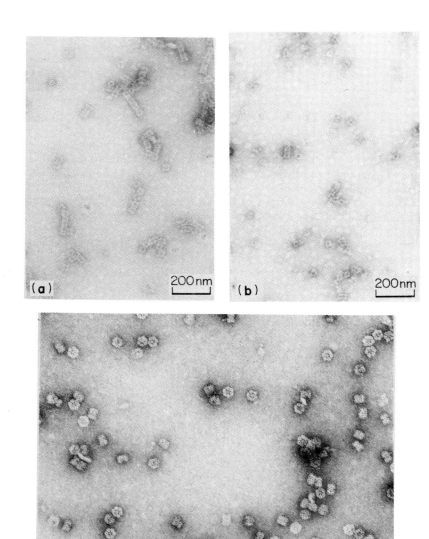

haemoglobin in the tissues. In arthropods, haemocyanin has been found in the Malacostraca (but not in other Crustacea), Xiphosura, and some Arachnida while it is absent from the Myriapoda and Insecta.

The copper remains divalent in oxyhaemocyanin and is bound directly to the protein rather than contained in a prosthetic group. The pigment is never found in cells but is dissolved in plasma as a high molecular weight polymer. In the Mollusca, these are large molecules of up to 9×10^6 daltons (see Fig. 3–1) but dissociation into smaller subunits occurs upon oxygenation.

3.3.3   Haemerythrin

Haemerythrin is found in corpuscles in a number of unrelated marine invertebrates: the Sipuncula, some Brachiopoda and Priapulida, and the single polychaete genus *Magelona*. It is a pigment in which iron is attached directly to the protein, rather than in haem. The active oxygen-binding site contains two ferrous atoms when the pigment is de-oxygenated but on reaction with oxygen, both atoms enter the ferric state, each donating one electron to an oxygen molecule to form a peroxide ion $O_2{}^{2-}$. A number of subunits combine to form octomeric molecules with a molecular weight of about 108 000 daltons.

(amino acids) . . . . . $Fe^{2+} - OH_2$ $H_2O - Fe^{2+}$ . . . . . (amino acids)

Deoxyhaemerythrin

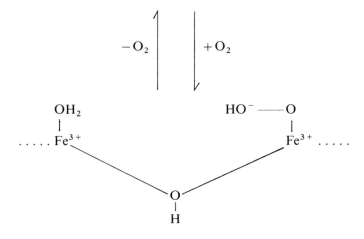

Oxyhaemerythrin

3.4 Oxygen affinity and the oxygen equilibrium curve

The function of a respiratory pigment is to carry oxygen to the tissues and carbon dioxide back to the respiratory exchange surface to be released into the environment. Three factors affect the amounts of gas which the pigment can deliver in a given time to its respective destinations:

(i) The concentration of respiratory pigment.
(ii) The oxygen affinity.
(iii) The blood flow through the tissues.

The way in which concentration and blood flow affects gas transport is self evident; for haemoglobin, 1 g of haemoglobin combines with 1.34 cm³ oxygen when measured on a haem basis. Thus the *oxygen carrying capacity* defines the amount of oxygen which can be carried by the pigment in the blood. Of course, it is not sufficient to have enough oxygen carrying capacity because the *oxygen affinity* of the pigment must be such that all haemoglobin molecules will pick up oxygen at the uptake surface and release a large proportion as they pass through the tissues. One way of assessing the respiratory role of a pigment is to place an organism in an environment containing carbon monoxide. Because pigments usually have a much higher affinity for monoxide, the oxygenation reaction may be rendered inactive. Comparisons of respiratory rates of CO-treated and control animals give an indication of the contribution to gas transport by the pigment.

Suppose that we take a solution of haemoglobin composed of monomeric subunits, having one haem per functional unit, and deoxygenate a few cm³ of the solution in a large flask using a vacuum pump. When all the oxygen has been removed the deoxygenated solution will be a blue or purple hue. If a little air is introduced, some of the oxygen combines with some of the deoxyhaemoglobin to form oxyhaemoglobin which is scarlet. Known amounts of air may be added until all of the haemoglobin is in the oxygenated form. If the results of this experiment are plotted on a graph with the partial pressure of oxygen on the horizontal axis and the percentage of oxyhaemoglobin on the vertical, the graph has the shape of a rectangular hyperbola (Fig. 3–2a). It rises steeply at first when all the monomers are free and it flattens out at the end when the free molecules are so scarce that only a high pressure of oxygen can saturate them.

Alternatively, this equilibrium curve (sometimes imprecisely called an "oxygen dissociation curve") may be represented by the graph in Fig. 3–2b in which the logarithm of the ratio of oxyhaemoglobin (Y) to deoxyhaemoglobin molecules (100–Y) is plotted against the logarithm of the partial pressure of oxygen. The hyperbola now becomes a straight

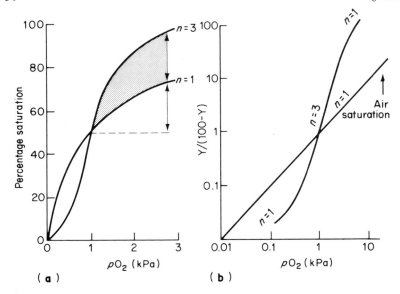

Fig. 3-2 Hypothetical oxygen equilibrium curves. (a) Sigmoidal shape for a pigment having cooperativity compared with the hyperbolic shape of one lacking cooperativity. (b) Logarithmic transformation of data from (a) to show Hill plots in which the slopes of the lines quantify the degree of cooperativity. Notice that the lines have been drawn for pigments with the same P_{50} and that a slope of $n > 1$ indicates cooperativity.

line sloping 45 degrees to the horizontal. The intercept of the line with the horizontal axis drawn at $Y/(100-Y) = 1$ gives the equilibrium constant P_{50}. This constant is the pO_2 at which exactly half of the haemoglobin is oxygenated and defines the *oxygen affinity* of blood. The greater the oxygen affinity of a pigment, the lower the pO_2 needed to achieve half-saturation and the smaller the P_{50}. In both the curve and the transformation, the shape or the slope of the line remains unchanged but a higher oxygen affinity shifts the equilibrium to the left and a lower oxygen affinity shifts it to the right. Hyperbolic curves are characteristic of all monomeric and dimeric pigments and some polymeric ones.

Consider now, a polymeric pigment such as the haemoglobin of an earthworm, or the haemocyanin of a garden snail, and an entirely different result is obtained. The equilibrium curve rises gently at first, then steepens and finally flattens out as it approaches saturation. This sigmoidal curve signifies that the pigment molecules are reluctant at first to take up oxygen, but that the affinity for oxygen increases with increased oxygen pressures. This phenomenon suggests that there is some kind of interaction between oxygen-combining centres and physiologists have called it haem-haem interactions, or more generally, to include non-haem proteins, *cooperativity*. In the logarithmic transformation, the curve begins with a 45° slope because at first the chance of an oxygen coming into contact with haem is low and each haem has an equal chance of reacting with oxygen. With a higher pressure of oxygen, the combining centres interact and the curve steepens. The steepness of a curve is defined by the tangent to its maximum slope and is known as Hill's coefficient, n, after the physiologist A. V. Hill who first attempted a mathematical analysis of the oxygen equilibrium. Without interacting oxygen-combining centres, a respiratory pigment has an n value of 1.0, but some curves are so steep that values of up to $n = 6$ have been obtained. Cooperativity in respiratory pigments is usually constant under physiological conditions and occurs only when the subunits are in aggregates of four or more. However, large aggregates do not necessarily show cooperativity.

3.4.1   Physiological significance of 'n'

Cooperativity has the functional effect of – (1) increasing the pO_2 diffusion gradient and therefore increasing the turnover of oxygen to the tissues, (2) increasing the volume of bound oxygen at the uptake sites and (3) decreasing the amount of pigment required to transport a given volume of gas. A sigmoidal curve is most obviously advantageous in a situation where the pigment can be fully saturated with oxygen at the uptake surfaces and unloaded when the tissue pO_2 coincides with the steep part of the curve. Such a pigment would be disadvantageous if the ambient pO_2 fluctuated to the extent that full saturation of the pigment could not be achieved. Moreover, it is less well adapted for storage functions when compared with a non-cooperative pigment because with a falling pO_2, most of the cooperatively bound oxygen will be released over a narrow interval of time.

On the other hand, non-cooperative pigments may be advantageous in an environment with greatly fluctuating pO_2, or when there is a need for an oxygen store. Thus for a given decrement in pO_2, or when the supply of oxygen is suddenly cut off, the release of bound oxygen will be similar in amount for any decrement over a wide range of pO_2.

3.4.2   Physiological significance of P_{50}

In contrast with n values, P_{50} may be altered *in vivo* by a number of important factors:
(i) Temperature
Like any other thermodynamic constant, P_{50} is sensitive to temperature because of the exothermic nature of the reaction between a

respiratory pigment and oxygen. With increasing temperature, oxygen affinity decreases according to the relationship,

$$\Delta H^\circ = 0.0192 \left[\frac{T^1 T^2}{T^2 - T^1} \right] \cdot \log (P_{50}{}^1 / P_{50}{}^2) \, kJ \, mol^{-1}$$

where ΔH° is the heat of oxygenation, T^1 and T^2 are the lower and higher temperatures in K, and $P_{50}{}^1$ and $P_{50}{}^2$ are the half-saturation values at T^1 and T^2 respectively. The values for the respiratory pigments generally fall within the range -40 to $-65 \, kJ \, mol^{-1}$. A low value of ΔH° implies a more stable molecular configuration in the transition from deoxy to oxy states than does a higher value. Physiologically these values are important because with a rise in temperature, the metabolic rate of an ectotherm increases and therefore the demand for oxygen is greater, thus more oxygen may be delivered to actively respiring tissue with a decrease in oxygen affinity.

(ii) pH

The pH scale is really a convenient way of expressing the quantity of hydrogen ions in solution as the negative logarithm of their concentration. Hydrogen ions have a pronounced effect on P_{50} for many haemoglobins, haemocyanins, chlorocruorins, and some haemerythrins. This effect, called the Bohr effect after the distinguished physiologist C. Bohr, is quantified by ϕ, and $\phi = \Delta \log P_{50} / \Delta pH$. Where $\phi = 0$, the oxygen equilibrium is independent of pH in the physiological range and there is no Bohr effect. Many haemoglobins, chlorocruorins and arthropod and cephalopod haemocyanins show a normal, or negative, Bohr effect so that a decrease in oxygen affinity is observed with a fall in pH. Values of -1.0 have been reported for very sensitive pigments. In contrast, a reverse, or positive, Bohr effect is seen in gastropod haemocyanins.

The physiological significance of a normal Bohr effect relates to the dissociation of carbon dioxide in solution according to the simple equation:

$$CO_2 + H_2O \rightleftharpoons H_2CO_3 \rightleftharpoons H^+ + HCO_3{}^-$$

Clearly, if hydrogen ions produced in the tissues by respiration are mopped up by the pigment, then more carbon dioxide can be dissociated and transported by the blood in bicarbonate form to be released to the environment. Moreover, in binding to the pigment, the hydrogen ions distort the shape of the pigment molecules causing oxygen to be released more readily by the decrease in affinity. Thus there is a reciprocal action between oxygen delivered and carbon dioxide removed and gas transport during a high level of activity will be assisted by a normal Bohr shift. The reason for a reverse Bohr effect is less obvious, but it does appear that many gastropod molluscs live in environments of high ambient CO_2

and have low levels of activity. Under these circumstances, a normal Bohr effect would hinder the uptake of oxygen, but the reverse effect secures saturation at the respiratory surface while having a minimal effect on oxygen delivery to the tissues.

(iii) Carbon dioxide

Some haemoglobins and haemocyanins bind molecular carbon dioxide although quantitatively this has less impact on the amount of gas transported. This phenomenon is distinct from the Bohr effect and carbon dioxide in molecular form is bound directly to the pigment when in the deoxy form and is released at the respiratory surfaces where oxygen is bound. This so-called 'carbamate' effect usually accounts for only a small fraction of carbon dioxide removed but it is likely to be of greater significance in invertebrates living in acidic habitats.

(iv) Salts

A variety of salts when in solution with a respiratory pigment, influence oxygen affinity by binding to the protein and changing its quaternary structure. In all invertebrate haemoglobins and crustacean haemocyanins, a decrease in salt concentration decreases the oxygen affinity of the pigment. In crabs and other invertebrates with haemocyanin in estuarine habitats the *in vivo* oxygenation of the body fluids is apparently unaltered upon dilution because the decrease in oxygen affinity opposes the rise in blood pH caused by the salts which shifts the curve to the left. This phenomenon may be of importance in these organisms subjected to regular fluctuations in the ionic composition of their blood by environmental factors since they are able to maintain a regulated supply of oxygen to the tissues. Some estuarine polychaetes increase their respiratory rates in low salinity and presumably a decrease in oxygen affinity secures more oxygen to meet this demand.

Gastropod haemocyanin behaves in the opposite manner: salts decrease oxygen affinity and reverse the Bohr effect. Although the end result of altered salinity may be the same, the significance of this phenomenon is poorly understood. Many pulmonate gastropods inhabit temporary bodies of fresh water or are fully terrestrial, and here the salt effect may serve to regulate oxygen delivery in dehydrating conditions.

Some inorganic ions appear to be important in modifying P_{50}. These are Cl^-, Mg^{2+}, Ca^{2+}, and PO_4^{2-}. In addition, some organic polyphosphates like adenosine triphosphate, have a more potent effect on P_{50} and may be thought of as 'supersalts'.

3.5　Adaptations to environment

A large number of recent investigations into the oxygen-combining properties of invertebrates respiratory pigments has led to the view that

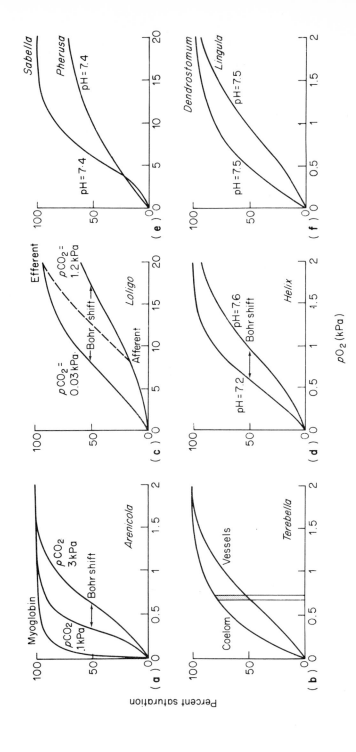

Fig. 3-3 (a) Oxygen equilibrium curves of whole blood from *Arenicola* equilibrated at two pCO_2s to illustrate the normal Bohr shift when hydrogen ions are released from the hydration of carbon dioxide. When carbon dioxide levels rise, the oxygen affinity of the haemoglobin is decreased and more oxygen can be released from the pigment. The curve for muscle myoglobin has a hyperbolic shape and does not shift in position with changes in pCO_2. Its high affinity assists the transfer of oxygen from the blood to the muscles. (b) During high tide oxygen is transferred from haemoglobin at a given pO_2 in the vessels of *Terebella*, to haemoglobin in the coelomocytes by an amount proportional to the shaded bar. When the tide recedes and the pO_2 declines, the large mass of coelomic haemoglobin will gradually release its store of oxygen. (c) The Bohr shift in *Loligo* haemocyanin is very large indeed. Efferent blood arriving in the gills has a higher affinity for oxygen than does afferent blood which has released its oxygen and combined with hydrogen ions on its return to the gills. The effective *in vivo* state of oxygenation is indicated by the pecked line. (d) The effect of pH on the position of the equilibrium curves of *Helix* haemocyanin illustrates the reverse Bohr effect in which a decrease in pH caused by carbon dioxide actually increases the affinity of the pigment for oxygen. Perhaps this is an adaptation which secures adequate oxygenation in its unventilated lung where the pCO_2 is high. (e) Oxygen equilibrium curves of chlorocruorin from the blood of the polychaetes *Sabella* and *Pherusa*, showing the marked contrast in both sigmoidicity and position of the curves. (f) Oxygen equilibrium curves for haemerythrin from *Dendrostomum* and *Lingula* which have different oxygen affinities and Hill coefficients at constant pH.

these properties are adaptive in serving to balance an organism's metabolic demands with the sources of oxygen in the near-environment. Our knowledge of oxygen uptake rates for any invertebrate and the physical parameters of its habitat is far from complete and values for these factors sometimes have to be assumed. The examples which follow have been chosen to illustrate just a few of the adaptive possibilities in invertebrates. Most of the examples are from marine habitats because this environment exhibits the greatest diversity in pigment function. Moreover, two of the four pigments – chlorocruorin and haemerythrin – are found only in marine invertebrates.

3.5.1 Haemoglobin in two marine worms

The lugworm *Arenicola marina* is often found burrowing in fine sediments on the shores of estuaries where interstitial water is deficient in oxygen. The oxygen deficiency is partly compensated when the worm irrigates its burrow and thus gains access to surface water of higher oxygen content. When the tide recedes the worm is abruptly cut off from this supply of oxygen for up to five hours. During these regular periods of hypoxia the pO_2 in the burrows falls from 20 kPa at the start of emersion to as low as 2 kPa before the incoming tide floods the burrows.

The oxygen-carrying capacity of the vascular blood in *Arenicola* is high (about 3.5 mmol dm^{-3}), but its small blood volume is insufficient to act as an oxygen store and sustain a routine rate of respiration for more than a few minutes at low tide. The oxygen equilibrium curve of the erythrocruorin contained in the blood is sigmoidal in shape ($n = 5$) and has a high oxygen affinity (Fig. 3–3a). Furthermore, the pigment has a marked Bohr effect ($\phi = -0.6$), an increased oxygen affinity in the presence of salts, and the equilibrium is relatively insensitive to temperature ($\Delta H° = -9.2\,kJ\,mol^{-1}$). Myoglobin in the body wall muscles has a much greater affinity for oxygen, a hyperbolic equilibrium curve ($n = 1$), and lacks a Bohr effect.

We may now enquire into the manner in which these properties of haemoglobin may enable *Arenicola* to function in its natural habitat. An estuarine environment is typically an unstable one with regular fluctuations in salinity, temperature, and oxygen concentration. The demand for oxygen increases with a fall in salinity (see section 4.6) and hence a rightward shift in the equilibrium curve may be seen as an adaptation which secures a greater oxygen delivery under diluting conditions. The low heat of oxygenation in *Arenicola* erythrocruorin may represent an eco-physiological adaptation in the sense that it safeguards the high oxygen affinity when, especially on sunny days, extreme high temperatures occur in shallow water covering the burrows. Under the low pO_2 conditions to which *Arenicola* is exposed at low tide, the high normal Bohr effect would be of survival value because it increases oxygen delivery to the tissues. At low pO_2, hydrogen ions may derive not only from metabolic carbon dioxide, but from additional sources of hydrogen produced by the acidic end-products of glycolysis (see Chapter 6).

The sigmoidal equilibrium curve is very steep and is an adaptation for oxygen delivery over a narrowly prescribed increment in pO_2. Thus when burrow water is at its lowest pO_2 (*ca.* 2 kPa) the erythrocruorin will still leave the gills fully saturated to deliver its load of oxygen on the steep portion of the curve. Transfer of oxygen from the blood to the myoglobin recharges the muscles with a store of oxygen between bursts of irrigation. Finally, the presence of the enzyme carbonic anhydrase in the blood will assist the rate of gas transport by catalysing the reaction between carbon dioxide and water to make available more hydrogen for the Bohr shift and bicarbonate ions for transport to the gills where the process is reversed: carbon dioxide is released and the affinity of the blood for oxygen is increased.

Terebella lapidaria, another intertidal polychaete, may be found inhabiting crevices in the mid-littoral zone of rocky shores. The micro-environment of this worm is more stable with regard to temperature and salinity than that of *Arenicola*. In addition to a discrete vascular system containing erythrocruorin, the coelom contains abundant erythrocytes charged with haemoglobin in sufficient quantity to give the worm a brick-red colour. When covered by the tide, the worm irrigates its burrow intermittently and its long red tentacles are well-placed for gas exchange. At low tide it cannot do so, and it is left quiescent in a thin film of water or in moist air for up to four hours.

The oxygen equilibrium properties of *Terebella* blood, apart from the effect of temperature, are essentially similar to those observed in *Arenicola*. In marked contrast, the equilibrium curve of the coelomic haemoglobin is hyperbolic and there is no Bohr effect (Fig. 3–3b).

When the tide recedes, the gills collapse and the contribution of the vascular blood must be negligible. But upon emersion, the coelomic haemoglobin with its higher oxygen affinity is fully saturated with oxygen. Since the coelomic fluid has a high oxygen capacity and occupies about 40 % by volume of a worm, then its contained haemoglobin could serve as a store of oxygen. In fact, as the internal pO_2 of the worm falls the coelomic haemoglobin releases oxygen to the tissues in approximately equal volumes per unit change in pO_2. Calculations from the capacity of the coelomic haemoglobin and observed respiratory rates in declining ambient oxygen tensions indicate that this store can maintain the animal in aerobic respiration throughout a low tide cycle. Aerial gas exchange may be supplemented by diffusion across the very thin dorsal body wall which has an area of 120 mm^2 and thickness of 0.15 mm in a 0.3 g worm. Using Krogh's diffusion coefficient for muscle, diffusion can account for up to half of the worm's oxygen under these conditions.

3.5.2   Haemocyanin in two molluscs

The terrestrial gastropod *Helix* and the marine cephalopod *Loligo* both have concentrations of dissolved haemocyanin that provide oxygen capacities of about 1 mmol dm^{-3} when the blood is saturated with air, about twice the concentration in the plasma alone. *Helix* is a typical land snail in having a vascularized mantle functioning as a simple diffusion lung without regular ventilation. As in other terrestrial animals, the pCO_2 of *Helix* blood tends to be high and contrasts with a low pCO_2 in *Loligo* because of the high solubility of carbon dioxide in sea water. Gastropods are slow moving invertebrates possessing an open circulation consisting of sinuses and large vessels. Cephalopods have a much greater scope for activity and have a muscular mantle which rapidly circulates the blood through a well-defined system of afferent and efferent vessels.

The haemocyanin–oxygen equilibria in *Helix* and *Loligo* are both sigmoidal, but the oxygen affinity of *Loligo* blood (see Fig. 3–3c) is much lower by comparison. This low affinity is further decreased by a fall in pH due to a large Bohr effect ($\phi = -1.8$). However, in *Helix* (Fig. 3–3d) and other gastropods, the haemocyanin has a remarkable property in that a fall in pH shifts the equilibrium to the left ($\phi = +0.4$). This reverse Bohr effect serves to increase the oxygen affinity of haemocyanin in the presence of carbon dioxide.

Are there environmental and metabolic correlates with these properties? In the well-oxygenated marine environment carbon dioxide produced by an active animal is readily eliminated by virtue of its high capacitance in sea water. Therefore, a normal Bohr effect in *Loligo* secures an increase in oxygen delivery when the call is for a greater oxygen demand. If the Bohr coefficient seems large, then it should be mentioned that the difference in pH between afferent–efferent blood in aquatic organisms is much less than the venous–arterial difference in terrestrial animals. In fact, the difference is only about 0.13 pH units in *Loligo*.

The less active terrestrial gastropod cannot ventilate its simple lung and the removal of carbon dioxide is therefore diffusion limited and the pCO_2 in the lung is higher than the value in air. Consequently, a normal Bohr effect would be disadvantageous in *Helix* because oxygen uptake in the lung would be impaired by a rightward shift in the equilibrium curve. A reverse Bohr effect on the other hand, is of adaptive value in *Helix* in that it ensures an adequate uptake of oxygen in the lung. Provided that the scope for activity is not very great, then the reverse Bohr shift would be of little consequence in unloading to the tissues.

3.5.3 Chlorocruorin in two marine polychaetes

The peacock worm *Sabella pavonina* is a tubicolous polychaete which lives subtidally anchored within a firm substratum. From the tube a crown of tentacles extends into a current of oxygenated seawater to serve as a respiratory and food gathering organ. *Pherusa plumosa* is an intertidal polychaete found in fine sediment under boulders where the supply of oxygen is irregular. Both species have vascular systems containing high concentrations ($ca.$ 5 mmol dm^{-3}) of the haemoprotein chlorocruorin.

Inspection of the chlorocruorin–oxygen equilibrium curves in Fig. 3–3e shows a sigmoidal curve for *Sabella* ($n = 5$) and a hyperbolic one for *Pherusa* ($n = 1$). The low n value in *Pherusa* would permit significant oxygen transport during a period of declining oxygen tension at low tide. A striking feature of all chlorocruorins is their large Bohr effect ($\phi = -1.0$) and low oxygen affinity. Perhaps these properties relate to the unusual haem structure and are not open to the adaptive possibilities shown by haemoglobin.

The Bohr shift of chlorocruorin is reminiscent of that in cephalopod haemocyanin and it is perhaps significant in this respect that worms with chlorocruorin also have vascular systems with clearly defined afferent and efferent vessels to the gills. If the Bohr effect is an adaptation that enhances oxygen unloading and carbon dioxide transport, then it will operate only when the pCO_2 of the habitat is low and when there is a difference in pH between afferent and efferent blood. As we have seen in *Loligo* a very small difference in pH will promote a greater release of oxygen at low pO_2. In addition, because of the buffering capacity of seawater, the pCO_2 is generally low except in unirrigated tubes or during pauses in irrigation when the pH falls due to a rise in pCO_2.

There is a problem in trying to interpret the respiratory significance of chlorocruorin. The low oxygen affinity, particularly in the case of *Pherusa*, indicates that the blood will be only 65 per cent saturated when in equilibrium with air, and hence the blood, unless it is extremely alkaline, can never utilize its full carrying capacity. It would be most useful to have measurements of blood pH in these worms, or even some simple observations on the state of oxygenation in the tentacles of living animals at various ambient pO_2s.

3.5.4 Haemerythrin in a sipunculid and in a brachiopod

All sipunculids (commonly called 'peanut worms') have a large volume of coelomic corpuscles and a small volume of vascular corpuscles containing haemerythrin. Little is known about the mode of life of these unusual invertebrates which typically inhabit burrows in soft muddy shores between the tides. In addition to the general body surface, sipunculids have short tentacles which offer an area for gas exchange. When *Dendrostomum zostericolum* is kept in seawater with a pO_2 of 20 kPa, the coelomic fluid pO_2 is only 3 kPa indicating that the body surface presents a major barrier to oxygen diffusion. Both coelomic and vascular haemerythrins in *Dendrostomum* have hyperbolic equilibrium curves (Fig. 3–3f) and lack a Bohr effect. The lower oxygen affinity of the vascular haemerythrin is consistent with a role whereby oxygen is transferred from it to the coelomic pigment where it awaits low tidal conditions to discharge its load of oxygen. Oxygen diffusion within the animal is facilitated by coelomic haemerythrin, a phenomenon which may partly compensate for poor circulation of the coelomic contents. The role of haemerythrin in *Dendrostomum* may be not unlike that described for the haemoglobin in *Terebella* which serves as an oxygen store for the low tide period.

A few brachiopods are known to contain haemerythrin in corpuscles in the body fluid. One of these lamp shells is *Lingula unguis*, a subtidal brachiopod which lives attached to rocks in currents of oxygenated water. *Lingula* haemerythrin (unlike that in sipunculids) has a sigmoidal equilibrium curve ($n = 2$, Fig. 3–3f) and a normal Bohr shift. Its function may parallel that of chlorocruorin in *Sabella*. Thus haemerythrin shows a variety of adaptive properties similar to the range observed in haemoglobin.

3.6   Evolution of the respiratory pigments

The sporadic distribution of the respiratory pigments is a phylogenetic puzzle. Why for example, do some polychaetes have no respiratory pigment, one species haemerythrin, a few chlorocruorin, some chlorocruorin mixed with haemoglobin, and the majority haemoglobin? And why is it that haemerythrin occurs in one polychaete, a few brachiopods and all sipuncluids?

By comparing the amino-acid sequences in these proteins we are able to speculate on which ones are progressive in evolutionary history and on the relationships between the phyla. With the diversity in respiratory habitats and the large variations in oxygen demand that are to be found in invertebrates, we would expect the pigments to be progressive.

The presence of haemoglobin in representatives from nearly all of the invertebrate phyla, the vertebrates, and even some plants, has prompted many biologists to suggest that haemoglobin may have arisen many times during evolution. This speculation is based on the observation that porphyrin, the precursor of haem, is found in all cells. The polyphyletic hypothesis has fallen from favour now that we have amino-acid sequencing data from the respiratory proteins of a number of different invertebrates. The sequence in intracellular haemoglobin from a worm or from a man differs in detail but not in the overall shape and size of the monomers. The extracellular haemoglobins, or erythrocruorins, and chlorocruorins stand in marked contrast. Their amino-acid sequences are entirely different, of different sizes, and comprise two peptide chains for each haem. The term erythrocruorin emphasizes the distinctive character of the extracellular haemoglobins although the term haemoglobin is still a useful collective term. The recent view favours the evolution of haemoglobin on two occasions, with the cellular pigment being an ancient relic of the earliest metazoans or even protozoans. The erythrocruorins and its variant chlorocruorin, may have arisen early in the evolution of the annelids. However, it is difficult to interpret the presence of erythrocruorin in the single gastropod species, *Planorbis*.

Haemocyanins from the Mollusca and Arthropoda are built according to entirely different plans. Arthropod haemocyanin has a molecular weight of less than 1×10^6 daltons while molluscan haemocyanins attain dimensions of up to 9×10^6 daltons. On the basis of their molecular architecture and amino-acid compositions, it has been suggested that the two haemocyanins may not be homologous, but rather the result of an independent origin and subsequent parallel or convergent evolution.

Insufficient observations on the structure of haemerythrin do not permit us to speculate on the distribution of one of the most interesting and rare pigments.

4 Factors of Energy Expenditure

In all probability life evolved in the absence of oxygen. This may well have been necessary because had free oxygen been available, the almost certain fate of the first forms of life would have been degradation by combustion. Yet one of the main problems which animals have to face is that of obtaining a sufficient supply of oxygen from the environment in order to check the loss of energy which tends to degrade all life according to thermodynamic principles. It is the purpose of Chapter 6 to examine the anaerobic process of energy production and how animals are adapted to oxygen-deficient environments.

While all animals modify their environments to some extent, it is the invertebrates which are to be found in the most diverse habitats and it is therefore hardly surprising that enormous variability is encountered in energy expenditure. Organismic as well as environmental factors influence the expenditure of energy and hence oxygen uptake in animals.

It is obvious that both the nature and rate of gas exchange in invertebrates are likely to be highly variable and represent the product of an extremely complex environmental situation. The scientific literature records many examples of variability in oxygen consumption both within and between species. A butterfly in flight consumes oxygen at a rate of $100 \ ml \ O_2 \ g^{-1} \ hr^{-1}$, nearly two hundred times the rate when at rest. An earthworm at rest consumes a mere $0.065 \ ml \ O_2 \ g^{-1} \ hr^{-1}$! However, as will soon be explained these respiratory measurements are meaningful only for the particular conditions of measurement.

A further division might distinguish quantitative factors (activity, body size, exposure temperature, nutritive level) from qualitative factors which are related directly to the nature of the environment – air or water. Many studies have made valuable contributions to our understanding of the influence of quantitative factors on invertebrate respiration but it is only recently that biologists have concerned themselves with more than just a few of the possible parameters. Many of the endogenous and exogenous variables are interdependent and difficult to interpret because of the simultaneous influence of several different parameters. In the following sections we shall look at a number of important factors which influence energy expenditure and consider their interrelationships. The examples have been chosen from mainly marine intertidal invertebrates because, as a group, they possess adaptations to a number of interacting environmental factors.

4.1 Oxygen availability

The supply of oxygen for aerobic respiration is ultimately limited by air, of which oxygen comprises 20.95 %; the partial pressure of oxygen (pO_2) in air is 21 kPa at sea level. While terrestrial invertebrates are usually assured of an unrestricted supply of oxygen, it is the aquatic invertebrates which are more likely to encounter low oxygen conditions.

The terms 'oxyconformity' and 'oxyregulation' are often used to describe the relationship between oxygen uptake and environmental pO_2. Since an animal cannot regulate its oxygen uptake over the entire range of possible oxygen tensions in the environment, some degree of conformity is always evident. The two ideal graphical forms that describe perfect conformity and perfect regulation are shown in Fig. 4–1. The oxygen tensions which mark the upper and lower limits to regulation are referred to as 'critical' points. This qualitative model, although once a popular one, is too simplistic and fails to describe the responses of real animals. Perfect regulators are rare and most invertebrates show at least

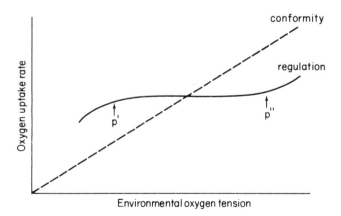

Fig. 4–1 Hypothetical relationships between rate of oxygen uptake and environmental oxygen tension. The two 'ideal' graphs show conformity (dashed line) where the rate is dependent upon external pO_2 over a wide range of tensions, and regulation (solid line) where the rate is independent of the environment over a range of pO_2 prescribed by critical oxygen tensions p' and p''. Beyond this range the rate conforms to external pO_2. Most animals actually show an intermediate response.

partial, or 'imperfect' regulation. A quantitative index would be useful for comparing and classifying the responses of different invertebrates and might obviate the need to force a continuously variable phenomenon into the dichotomy 'conformer' or 'regulator'. A number of mathematical models have been proposed in an attempt to describe the responses of animals to declining oxygen tensions and thus the degree of oxyregulation. One of the simplest models is linear regression which describes a straight line according to the equation

$$Y = B_0 + B_1 (\log X)$$

where Y is the rate of oxygen uptake, and X the partial pressure of oxygen. Since B_0 and B_1 are constants representing the Y-intercept and slope of the line respectively, it will be apparent that the constant B_1 might be a useful index of oxyregulation. Although this model has the great advantage of familiarity, it cannot accurately describe many of the responses by real animals such as those shown for a range of marine invertebrates in Fig. 4–2. One recent model which, though by no means perfect, adequately accommodates many observations. This model is a quadratic equation that is more powerful than linear regression because it incorporates an additional unknown variable. In the equation

$$Y = B_0 + B_1 X + B_2 X^2$$

the coefficient B_2 is an informative index of the shape of the curves in Fig. 4–2 as may be seen in Table 5. A value of B_2 which is not significantly different from zero implies oxyconformity, while the highest values reflect considerable regulatory abilities. Thus the degree of regulation is reflected visually in the degree of curvature of the graphs in Fig. 4–2 where the scyphozoan *Aurelia* is a perfect conformer while the mussel *Modiolus* is a very good regulator.

Unfortunately, the analyses from this, or any other regression model, have not suggested an ecological pattern of responses to environmental oxygen. One might expect that the polychaete worm *Capitella*, sometimes used as an indicator of pollution and living in low oxygen sediments, should show some degree of regulation while at the other extreme, the starfish, *Asterias* living in a well-oxygenated surf zone might be expected to be a conformer. However, Table 5 and Fig. 4–2 show that the reverse is the case!

There is recent evidence which suggests that a variety of endogenous and exogenous factors influence the degree of oxyregulation. For example, the bivalve *Rangia* is not influenced by previous exposures to low oxygen and the index remains constant. On the other hand, the sea cucumber *Thyone* tends towards conformity following prior hypoxic exposure. A similar phenomenon has been noted in leeches and chironomid larvae living in freshwater habitats. Body size appears to be a

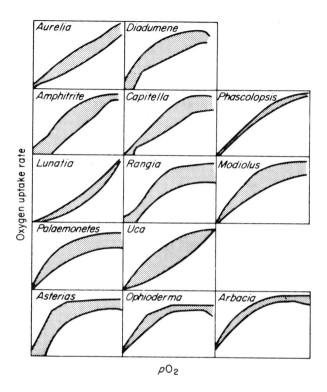

Fig. 4–2 Graphs of actual relationships between oxygen uptake rates and environmental oxygen tensions for a variety of marine invertebrates. The stippled area in each graph represents the range of responses shown by a particular organism and the degree of oxyregulation is reflected visually in the degree of curvature of the graphs. A linear graph indicates conformity, while a hyperbolic graph indicates regulation at higher environmental tensions. (Based on Mangum, C. P. and van Winkle, W. (1973). *Amer. Zool.*, **13**, 529.)

further factor to be considered in the mussel *Mytilus edulis*. Large *Mytilus* exhibit a high degree of regulation in hypoxic situations while small ones tend to conformity. Respiratory independence increases markedly with increasing body size in several other bivalves. The physiological condition of the animal is also known to affect the regulatory ability of mussels. *Mytilus* which have recently been brought in to the laboratory are able to regulate their oxygen uptake although this ability is rapidly lost with starvation. There is also evidence that the

Table 5 The degree of oxyregulation shown for a number of marine invertebrates represented by various phyla. Invertebrates with infaunal habits might be expected to live in conditions of lower environmental oxygen tensions than those with epifaunal habits. Regulation is quantified by the coefficient, B_2, and a more negative value of B_2 implies a greater degree of oxyregulation than does a more positive value. (From Mangum, C. P. and van Winkle, W. (1973). *Amer. Zool.*, **13**, 529.)

Marine invertebrate	Habit	$B_2 (\times 10^3)$
Cnidaria		
Scyphozoa		
Aurelia aurita	epifaunal	−0.007
Anthozoa		
Diadumene leucolena	epifaunal	−0.041
Annelida		
Amphitrite ornata	infaunal	−0.040
Capitella capitata	infaunal	−0.008
Sipuncula		
Phascolopsis gouldi	infaunal	−0.036
Mollusca		
Gastropoda		
Lunatia heros	infaunal	+0.049
Bivalvia		
Modiolus demissus	infaunal	−0.228
Rangia cuneata	infaunal	−0.070
Arthropoda		
Palaemonetes pugio	epifaunal	−0.047
Uca pugilator	infaunal	−0.052
Echinodermata		
Asteroidea		
Asterias forbesi	epifaunal	−0.105
Ophiuroidea		
Ophioderma brevispina	epifaunal	−0.063
Echinoidea		
Arbacia punctulata	epifaunal	−0.047
Holothuroidea		
Thyone briareus	infaunal	−0.030

environmental factor of salinity has an effect on the degree of oxyregulation shown by bivalves.

If there is any generalization to be made it is that species at an elementary level of organization tend to show small departure from conformity while those at a higher level of organization indicate a greater constancy of respiratory rate. Since there is increasing evidence to indicate that both environmental and organismic factors affect the degree of oxyregulation, considerable caution ought to be exercised before classifying species as oxyconformers or oxyregulators.

4.2   Exposure temperature

4.2.1   The Q_{10} relation

The rate of any biochemical reaction is dependent on temperature. Therefore, it might be expected that the oxygen uptake in whole animals is also dependent upon exposure temperature. Metabolic rate responses to temperature are usually quantified by the Q_{10} relationship where,

$$Q_{10} = \left[\frac{R_1}{R_2} \right]^{10/(T_1 - T_2)}$$

where R_1 and R_2 are the rates of oxygen consumption measured at temperatures T_1 and T_2 in °C respectively. Thus, if a reaction proceeds at a constant rate, even though the temperature is raised, or lowered by 10°C, then $Q_{10} = 1.0$. For most biochemical reactions the Q_{10} is approximately 2 and this indicates a doubling of the respiratory rate with a 10°C rise in temperature.

The living organism is a vast manufactory of biochemical processes and many reactions proceed simultaneously. This is precisely the reason that temperature effects on the oxygen consumption of whole animals are exceedingly complex and difficult to interpret. Added to this is the difficulty of obtaining ecologically meaningful results of oxygen consumption for diurnal, tidal and seasonal fluctuations in environmental temperature. It is hardly surprising therefore, that Q_{10} values for oxygen consumption in whole animals often deviate from the magical number of 2 and values which range from minus numbers to over 70 have been reported!

4.2.2   Ectothermy and oxygen consumption

Invertebrates are generally classified as 'cold-blooded' species and contrast with 'warm-blooded' vertebrates for obvious reasons. The term *poikilothermic* ('poikilo' is a Greek word meaning 'varied') is often used to describe the thermal aspects of metabolism. There are invertebrates however, such as those of the deep sea infauna and antarctic seas, which live at constant temperatures and for this reason the term *ectotherm* is preferred as it denotes the source of body heat.

The influence of temperature on the respiration of invertebrates has received more attention than other environmental factors perhaps because it is an obvious and easily measured parameter which varies diurnally, seasonally and with latitude. These last two categories involve essentially long-term differences in environmental temperature and it is therefore useful to separate them from rapid diurnal fluctuations in the thermal regime. There are many patterns of variation in the metabolic response of invertebrates to temperature, and this reflects in part the different effects of temperature on factors such as activity level, body size, and nutritional state of the organism concerned.

4.2.3   Influence of activity

The first and perhaps the most important single factor which affects the temperature relationships of respiration is activity. The rate of activity, and consequently the active rate of respiration, nearly always increases logarithmically with temperature with a Q_{10} of approximately 2. Such processes include ciliary activity in mussels, cirral activity in barnacles, radular activity of gastropods and the heart rates of many invertebrates. It is not surprising to find therefore, that in most instances where the experimental animal is either fully active or showing a routine level of activity, that the respiration rate is essentially dependent upon temperature.

In contrast, a wide variety of different temperature effects have been recorded for the respiration of quiescent invertebrates. Such relationships vary from temperature dependence throughout the range of thermal tolerance of the organism concerned to rate-temperature curves which show low Q_{10} values over at least part of the environmental temperature range. The occurrence of such regions of temperature insensitivity appears to be confined to invertebrates subjected to regular cyclical temperature fluctuations such as are associated with the ebb and flow of the tides. In the intertidal sea urchin *Strongylocentrotus purpuratus* the metabolic rate is essentially independent of temperature between 12 and 21°C whereas the subtidal *S. fransiscanus* has a high value for the Q_{10} throughout its environmental temperature range.

4.2.4   Availability of food

Nutritional factors appear to influence the temperature dependence of respiration in a number of invertebrates. In the shore crab *Carcinus maenas* starvation not only suppresses respiration but metabolism becomes much less dependent on temperature. This suggests that temperature dependence of energy expenditure may be associated with the availability of food.

4.2.5   Significance of low Q_{10}

The occurrence of decreased rates of respiration, with their associated low Q_{10} values in a wide variety of invertebrates living in thermally unstable environments may thus represent a means by which energy is conserved despite a rise in temperature. Equally, the occurrence of low

Q_{10} values may reflect the presence of low metabolic substrates induced by starvation or metabolic stress.

4.2.6   Thermal acclimation

Factors which are principally associated with long term temperature change such as occur seasonally or with latitude may induce an organismic response. If an ectotherm is kept at an altered temperature for a number of days, its rate of oxygen uptake often shows some compensation, i.e. it becomes *acclimated*. If it is now returned to the original temperature the rate does not return directly to the original level but rather to a higher or lower value according to the direction of acclimation. A description of the possible acclimatory responses of invertebrates is given in Fig. 4–3. When animals have been held at different temperatures for several days and their rate-temperature curves are then found to be coincident (Pattern I) then there is no acclimation.

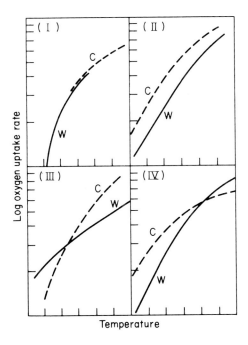

Fig. 4–3   Patterns of oxygen uptake rates at different temperatures for animals acclimated to warm temperatures (W) or cold temperatures (C). For further explanation refer to text. (Based on Prosser, C. L. (1973). *Comparative Animal Physiology*. Saunders.)

Examples of this pattern are to be found in most antarctic invertebrates whose previous thermal history is constant in sea water on the verge of freezing at $-1.9°C$. More frequently, however, there is a translation in the rate-temperature curve without a change in slope (Pattern II). Many invertebrates show this pattern of nearly complete compensation – *Planaria*, Coleoptera, and some Decapoda. Pattern III illustrates a third possibility where the rate-temperature curve is rotated following acclimation. Post-acclimated animals having a change in Q_{10}, or slope, show rotation. Commonly, acclimation is a combination of translation and rotation so that Q_{10} is reduced by cold exposure but some compensation is evident; an example of Pattern IV is the ventilation rate in *Mytilus*.

These patterns should not be considered fixed responses of organisms. A number of exogenous and endogenous factors are known to modify the acclimatory response and these include the temperature range of measurement, season, hormonal state, food availability, salinity and genetic capacity for temperature acclimation.

4.3   Activity level

Invertebrates display an enormous range of activity at both the individual level and for different species compared. Thus a butterfly in flight consumes 200 times more energy than when at rest, and a swimming squid expends a great deal more energy than a browsing mollusc. The term *basal* rate is used to quantify energy expenditure in vertebrates at rest. Unfortunately, there is no clearly observable basal rate in invertebrates and thus quantification of minimal energy expenditure is extremely difficult to assess. For this reason, the term *standard* rate is preferred when describing quiescent invertebrates. Given certain environmental circumstances, respiration may fall below the standard rate to levels at which the distinction between life and death is difficult to perceive. Such circumstances are regularly encountered by intertidal animals at low tide. With a cessation of heart and ciliary beats and other signs of overt activity, an animal may obtain a minimal supply of energy from anaerobiosis (see Chapter 6).

The standard rate of respiration is most easily quantified in those organisms which either ventilate their gas exchange surfaces, as in polychaetes and bivalves, or are mobile as are many arthropods. The graph in Fig. 4-4 shows the relationship between activity and oxygen consumption in the mussel *Mytilus edulis*. A maximum level of activity may be elicited under constant environmental conditions if the mussel is fed a suspension of its favourite algae. By extrapolation to zero activity, the standard rate is defined.

The difference between active and standard rates defines the *scope for*

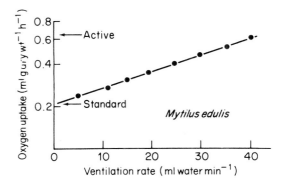

Fig. 4–4 Graph showing the relationship between ventilation rate (ml water min^{-1}) and oxygen uptake (ml O_2 g dry wt^{-1} hr^{-1}) in a 1.0 g specimen of *Mytilus edulis* at 15°C.

activity and this may vary according to the work performed during locomotory or irrigatory activity. The energetic cost of activity is difficult to estimate and may, for example, be as high in a relatively sluggish animal which, during locomotion moves a shell many times its own tissue weight, as in a pumping bivalve. Comparisons between the scope for activity for different invertebrates have, therefore, little quantitative significance unless expressed in terms of work performed during activity. The scope for activity may also vary according to a number of other exogenous variables of which temperature and nutritional factors are considered elsewhere in this chapter.

4.4 Body size

Body size is an important endogenous factor which affects the respiration of animals and it has been demonstrated for a variety of invertebrates that the line relating the logarithm of metabolism to the logarithm of body weight has a slope of approximately 0.75. Data relating oxygen consumption rates to body size have been compiled for all sorts of animals in Fig. 4–5.

However, there are instances where the slope of the regression line varies according to certain environmental factors such as temperature. These instances suggest that temperature or other external conditions may have a different effect on large and small individuals of the same species. One recent solution to this complex situation involves the technique of multiple regression analysis on a large number of observations of respiration rates under a variety of experimental conditions. The statistical model is then used to predict the effect of external factors

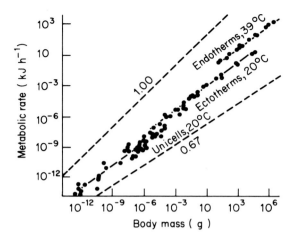

Fig. 4–5 The rates of oxygen uptake for a wide variety of organisms when calculated per unit body mass and plotted against body mass (log coordinates) tend to fall along regression lines with a slope of 0.75.

on metabolism. An analysis of this type for active and inactive marine snails under a variety of exposure temperatures is depicted in Fig. 4–6 where it may be noted that the rate-size relationship for a particular animal is not constant as might be supposed from Fig. 4–5.

4.5 Nutrition

It is undoubtably true that many invertebrates collected for laboratory experiments have been starved for some time, often days, before the measurements are made. It is now known that nutritional conditions greatly influence oxygen consumption rates. The effect of starvation on the oxygen consumption of the shrimp *Crangon vulgaris* is such that the decrease in respiration of active animals is greatest during the first few days of starvation but that the rate in quiescent animals is almost unchanged. In addition, small individuals decrease their oxygen consumption at a faster rate than do larger ones of the same species.

These decreases in respiratory rate may be attributed to changes in metabolism during depletion of an animal's energy reserves. In the barnacle *Balanus balanoides*, carbohydrate reserves are utilized initially, but after these have fallen to approximately 10% of the body weight, proteins and lipids are mobilized as metabolic substrates. Furthermore, the temperature coefficient for the moulting frequency of *Balanus* varies according to the substrate being utilized. The Q_{10} value is approximately

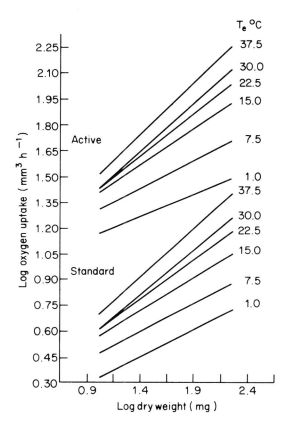

Fig. 4–6   Graphs showing the influence of short-term exposure temperature (T_e) on the relationship between metabolism and body size in the marine snail *Littorina littorea*. (Data from Newell, R. C. (1973). *American Zoologist*, **13**, 513.)

2.0 in well-fed barnacles using carbohydrate as a substrate but falls to 1.0 with protein and lipid utilization. Nutritional factors might be expected to have ecological significance in interpretation of seasonal cycles of energy expenditure for invertebrates subjected to marked fluctuations in food availability at different times of the year.

4.6   Salinity

It is commonly observed in marine invertebrates that respiratory rates increase with decreasing salinity of the surrounding sea water. One might expect that invertebrates living in estuarine areas of fluctuating salinity (termed *euryhaline*) might exhibit respiratory adaptations not seeñ in *stenohaline* animals living in water of nearly constant salinity. As predicted, stenohaline species fail to regulate oxygen consumption under low salinity but the euryhaline mussel *Mytilus edulis* recovers its regulatory capacity after a few hours. There is no generally accepted explanation of these findings. However, such effects should not be attributed to salinity-linked changes in oxygen concentration because the pO_2, and hence the rate of diffusion of oxygen remain the same.

4.7   Photoperiod and rhythms

Many invertebrates show oscillations in respiration which persist in laboratory conditions where the major entraining environmental cycles are precluded. These rhythms may have periods approximating a year, a synodic month (29.5 days), a lunar day (24.8 hours), a solar day, and a tidal interval (12.4 hours). It is thought that the timing of these rhythms is under the control of a mechanism within the organism called the 'biological clock'.

Photoperiod has a marked influence on energy expended by the crab *Pachygrapsus*. In an 8-hour photoperiod this crab consumes 55 % more oxygen than when exposed to light for 16 h every day. Bimodal tidal rhythms are often found in intertidal invertebrates. The fiddler crab *Uca pugilator* exhibits a predominant and persistent tidal rhythm which is lost after the crab's eyestalks have been amputated. Removal of the eyestalks destroys the tidal component of the rhythm (see Fig. 4–7) but has no influence on the diurnal rhythm. The implication here is that the effect of a physical factor on energy expenditure is mediated by a hormone.

The respiratory rhythm in crabs is in approximate phase synchrony with their activity rhythm, implying that oscillations in oxygen uptake are a consequence of activity. In *Uca* a lunar rhythm is superimposed on the circadian rhythm in an unexpected manner. Oxygen uptake is 50 % higher at 06.00 hr than at 18.00 hr, and in a lunar cycle the consumption is 30–50 per cent higher when the moon is at either its highest or lowest transit than when it is at the horizon. A most surprising discovery has been the influence of variations in the intensity of cosmic radiation on oxygen consumption in fiddler crabs. It is possible that many other animals might be influenced in the same way.

4.8   Conclusion

The response of a whole animal to a particular parameter is bound to be more complex and variable than the functional responses of any of its

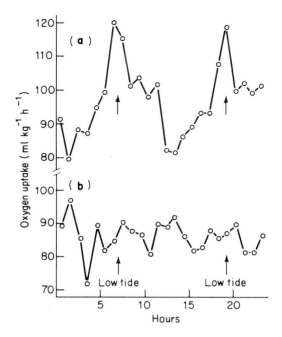

Fig. 4–7 The mean tidal rhythm of oxygen consumption in the fiddler crab *Uca pugilator* in dim constant light and 25° C (**a**). As seen in (**b**), the rhythm is lost after the crabs' eye-stalks have been amputated. (Based on Palmer, J. D. (1974). *Biological Clocks in Marine Organisms.* Wiley-Interscience.)

parts, for an animal is an integrated whole, and, unlike a thermodynamic system in equilibrium, it is a dynamic one. The living organism may be viewed as a system in which energy is selectively channelled into activity, growth, and reproduction; it is not a system in equilibrium.

5 Regulation of Respiration

The blood system delivers oxygen to tissues and eliminates carbon dioxide not only according to variations in the demand for energy, but also, as far as possible, in the face of fluctuating oxygen concentrations in the environment. We have already described the degree of oxyregulation in whole animals and now we must examine the mechanisms which contribute to respiratory homeostasis. Regulation in invertebrates is just as complex as regulation in vertebrates, especially mammals, because the latter have achieved a greater measure of independence from their environment. Invertebrate respiratory systems might therefore be expected to operate in spite of large fluctuations in the physico-chemical composition of their internal and external environments.

Invertebrate circulatory systems are classified as either 'open' or 'closed' although in some cases, the distinction cannot clearly be defined. In an open system the blood, called *haemolymph*, is circulated under low pressure from a pulsatile heart through vessels, into tissue spaces where it bathes cells, and then it is returned through sinuses to the heart. In a closed system blood is circulated under greater pressure through narrow vessels and capillaries to furnish tissues with oxygen before returning to the heart. In this case, the blood does not mix with interstitial fluid but flow is directed towards specific sites where oxygen can be unloaded.

All animals, other than strict oxyconformers, have a system of respiratory control ensuring adequate oxygenation of tissues. Moreover, carbon dioxide and hydrogen ions influence oxygen delivery and their control too must be accommodated within an overall strategy of respiratory regulation. Various aspects of regulation in many invertebrates have been studied, but physiologists have tended to focus on large species because of technological difficulties imposed by small size. In the remainder of this chapter we examine regulation in a terrestrial insect and an aquatic crustacean with open circulation, and a marine annelid having a closed circulation.

5.1 Regulation of tracheal respiration in an insect

Insects are unique in having a tracheal system which relieves the circulatory system of its primary respiratory function. Many of them rhythmically ventilate their airways thus accelerating the movement of gases to and from the tissues. In the desert locust *Schistocerca,* inspirations and expirations are normally effected by dorso-ventral abdominal movements but in times of respiratory stress, these actions may be supplemented by pumping movements of the head and thorax. The rate of ventilation is governed by a neural pacemaker located in the abdominal ganglion.

At rest, *Schistocerca* pumps $40 \, dm^3$ air kg body weight^{-1} hr^{-1}, but when inspired air is replaced by a breathing mixture containing 5 per cent carbon dioxide in air, the rate rises to approximately $250 \, dm^3 \, kg^{-1} \, hr^{-1}$ indicating that the ventilatory mechanism is sensitive to carbon dioxide. In addition, carbon dioxide has a direct effect on the closure muscles of the spiracles, causing them to relax and remain open, thus allowing an unrestricted flow of gas. It ought to be pointed out that the spiracles do not remain fully open when an insect is at rest because of the conflicting requirement to conserve water which might be lost by the exit of water vapour from the tracheae.

An additional supply of oxygen is needed to sustain an insect in flight. This is largely met by rapid ventilation through changes in thoracic volume caused by wing movements. The oscillations in intrathoracic pressure produced by these changes in volume induce flow, with air sucked in and then expelled with each wingstroke. Ventilation during flight is sometimes termed 'autoventilation' to distinguish it from active abdominal pumping when at rest.

All insects depend on liquid diffusion from the terminals of the tracheoles to actively respiring cells and regulation of the final diffusion distance occurs at the tracheolar endings. Muscular movements and a limited capacity for haemolymph circulation can shorten the diffusion pathway but the most remarkable and efficient mechanism for control of gaseous exchange was discovered by the insect physiologist V. B. Wigglesworth. He found that a short-term increase in oxygen demand was accompanied by withdrawal of the fluid which fills the terminal parts of the tracheoles into surrounding tissues. The water-binding power around the endings is influenced by changes in the osmotic pressure of the haemolymph in which they are bathed. During muscular contraction, especially if the oxygen supply is deficient, the osmotic pressure of the haemolymph rises due to an accumulation of acidic metabolites. This rise causes the now hypotonic fluid from the terminals to be drawn into adjacent extracellular fluid, and thus air extends into the finest branches of the tracheoles. When the muscles return to rest, the metabolites are removed as oxidized products of low osmolarity and fluid re-enters the tracheoles. Since gaseous diffusion is faster by several orders of magnitude than liquid diffusion, the advantage of bringing air closer to respiring tissue is apparent. When the demand for more oxygen is sustained, such as during growth, or with colonization of an oxygen-poor environment, then compensation occurs by expansion of the tracheal system into hypoxic tissues.

5.2   Regulation in a crustacean with open circulation

5.2.1   Water flow in the branchial chamber

The gills of the decapod Crustacea are suspended in a branchial chamber and ventilated by the pumping action of a pair of broad exopodites called *scaphognathites*. Each scaphognathite moves as a rigid paddle impelling a forward flow of water over the gill surfaces. Water is drawn by suction into the branchial chamber from inhalent apertures situated between the limb bases and is expelled through anterior excurrent channels opening just below the antennae. The anatomical arrangement of these structures in relation to the direction of water flow is represented in Fig. 5–1 by the freshwater crayfish *Procambarus clarki*.

Each beat of a scaphognathite produces an oscillation in pressure which can be monitored by a transducer implanted in the epibranchial chamber to produce a trace like the one figured. An occasional 'coughing' reflex briefly reverses the flow to clear debris and produces a positive pressure spike in the trace.

Because water contains less oxygen than air, aquatic animals must pump a large volume of water over their respiratory surfaces in order to obtain an adequate supply of oxygen and it might be hypothesized that in these animals, ventilation rates would be sensitive to changes in ambient oxygen tensions. This is indeed the case for a number of invertebrates including the lobster *Homarus*. At low oxygen tensions *Homarus* increases the frequency of its scaphognathite beats and hence water flow (Fig. 5–2). Experiments with *Homarus* have shown that the scaphognathite beat is insensitive to changes in the carbon dioxide tensions of the branchial chamber, but sensitive to water oxygenation. Below a critical pO_2 there occurs an *apnoeic* phase: that is, a phase of suspensed respiration. Regulation of water flow over a broad range of ambient pO_2 ensures a buffering of internal pO_2 and thus a respiratory homeostasis.

5.2.2   Oxygen transport by the blood

At normal ambient pO_2 when oxygen in post-branchial blood is high an animal is said to be in the *normoxic* state and a low diffusion resistance occurs across the gill surfaces. Blood flow within a gill filament of the crayfish is complex, at first following a concurrent, then a countercurrent course with respect to the direction of water flow ensuring that the blood is fully saturated with oxygen. The amount of this oxygen that is available to the tissues is about the same over a wide range of oxygen

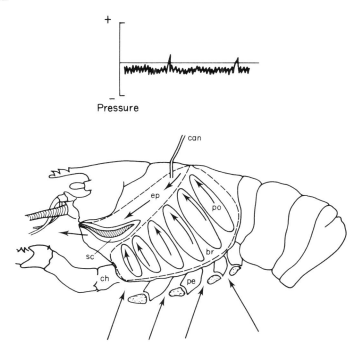

Pressure

Fig. 5–1   Direction of water flow within the branchial chamber of the freshwater crayfish *Procambarus*. Changes in hydrostatic pressure recorded from a cannula (can) inserted in the epibranchial space (ep) are show in the trace. Negative branchial pressures cause water to be drawn into the chamber but an occasional positive spike represents a "coughing" reflex which clears the chamber of debris br, branchiostegite; ch, cheliped; pe, pereiopod; po, podobranch; sc, scaphognathite.

tensions due to the contribution made by haemocyanin-bound oxygen. When *Homarus* is exposed to declining oxygen tensions (Fig. 5–3) the amount of oxygen physically dissolved in the blood becomes insufficient to sustain respiration and the amount of oxygen released by the haemocyanin becomes more important. This is so because not only does the haemocyanin have a higher carrying capacity than the plasma alone, but it releases the bulk of its bound oxygen when internal oxygen tensions are low, thus making use of the sigmoidal oxygen equilibrium curve. Tiny cavities, called *lacunae*, lying under the gill epithelium provide for the possibility of further regulation by shunting the blood in

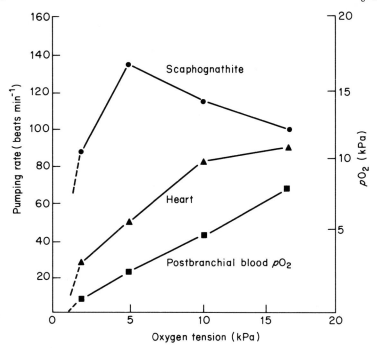

Fig. 5–2 Control of ventilation in the lobster *Homarus*, exposed to different external oxygen tensions. The scaphognathite, rather than the heart, increases its rate of beating in declining oxygen down to a critical tension of about 4 kPa, below which breathing is suspended and the animal is said to be in an apnoeic state. Note that if the ventilation rate did not increase down to 4 kPa, then the post-branchial tension might be considerably reduced. (Based on McMahon, B. R. and Wilkens, J. L. (1975). *J. exp. Biol.*, **62**, 637–55.)

times of stress. A hypoxic animal may shunt blood further along the length of its gill filaments to gain a maximum degree of oxygen saturation in the afferent blood supply. Conversely, when a crayfish is exposed to an abnormal osmotic environment, blood flowing through the gills may be protected from osmotic stress by shunting blood proximally across the base of the gills in order to minimize ionic exchanges with the unfavourable environment.

5.2.3 Acid-base regulation

Most of the carbon dioxide produced in cells by respiration is

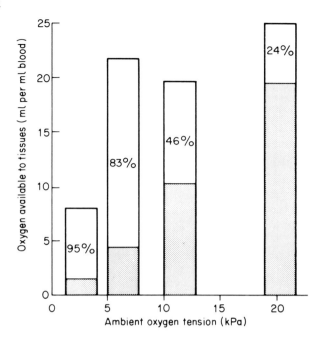

Fig. 5–3   Proportions of oxygen delivered to tissues by haemocyanin in the lobster, *Homarus* exposed to four ranges of oxygen tension. □ , from haemocyanin; ■ , from physical solution. The contribution made by haemocyanin-bound oxygen is more important than physically dissolved oxygen when the animal is in low oxygen tensions. (Based on McMahon, B. R. and Wilkens, J. L. (1975). *J. exp. Biol.*, **62**, 637–55.)

hydrated according to the reversible reaction

$$CO_2 + H_2O \rightleftharpoons H_2CO_3 \rightleftharpoons HCO_3^- + H^+$$

When blood reaches the gills, low ambient pCO_2 reverses the reaction and releases carbon dioxide into the environment. Both the rate of formation of reactants and products which is catalyzed by the enzyme carbonic anhydrase, and the concentration of the components at equilibrium have the potential to alter the acid-base status of the blood. The relationship at equilibrium is predicted by the Henderson-Hasselbalch equation,

$$pH = pK + \log \frac{[HCO_3^-]}{\alpha PCO_2}$$

where pK is the dissociation constant for carbonic acid and α is the solubility coefficient for carbon dioxide.

Under conditions of low ambient oxygen tensions, *Astacus* has an elevated ventilation rate and hence a faster than normal removal rate of carbon dioxide creating a *hypocapnic* (low pCO_2) condition in which the blood has a higher pH, thus said to be *alkalotic*. This condition ensures a higher affinity of the haemolymph for oxygen in the gills because of the haemocyanin's strong Bohr effect. Delivery of oxygen may therefore be regulated by a change in the acid-base status of the haemolymph in response to environmental oxygen tensions. The interactions between pH, bicarbonate, and carbon dioxide in this regulation may be expressed by the 'buffer line' which shows the relationship pH and bicarbonate concentration in the blood (Fig. 5–4).

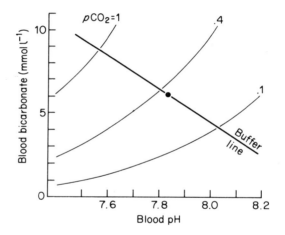

Fig. 5–4 Acid-base diagram of bicarbonate concentrations *versus* pH in the prebranchial haemolymph of the freshwater crayfish, *Astacus*. The solid point on the buffer line indicates the normoxic value and the curved lines are pCO_2 isopleths. Regulation in altered oxygen regimes involves a shift along the buffer line to a new acid-base status, passing down the slope with hypoxia and up with normoxic recovery.

5.3 Regulation in an annelid with closed circulation

5.3.1 Water flow in the burrow

The European lugworm *Arenicola marina*, lives in a sandy burrow between the tides where it periodically ventilates water through its burrow in a tail-to-head direction by means of piston-like waves that run along its dorsal body surface. The vigorously ventilated stream of water brings oxygen from the tail shaft and in passing over the gills of the worm, carbon dioxide is removed to water in the head shaft. Evidence of an oxygenated stream of water is clearly seen in the lighter zone of sand lining the burrows and contrasts with the dark reducing conditions of the surroundings.

A worm in its burrow follows a cyclical pattern of activity, first retreating towards the tail shaft to deposit faeces at the surface, and then moving to the bottom of the burrow while continuously pumping water in a forward direction. Water movements in a burrow may easily be traced on a kymograph in the laboratory using a float in the tail shaft of an artificial burrow as illustrated in Fig. 5–5. In this way the worm's activities can be recorded continuously for many days. Decerebrate animals show a similar pattern of activity which suggests that ventilation is not under the control of a pacemaker in the cerebral ganglion.

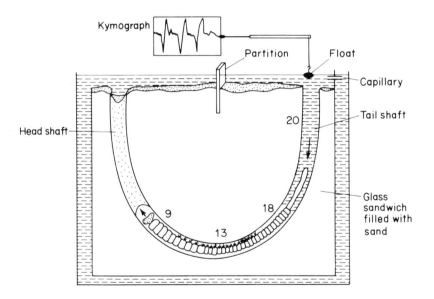

Fig. 5–5   Diagram showing an apparatus used for recording ventilatory patterns in the burrow of the lugworm, *Abarenicola affinis*. Heavy arrows indicate the direction of water flow and the numbers the pO_2 (kPa) of water sampled at various points along the burrow. A float placed in the tail chamber of the burrow is attached to the writing arm of a kymograph drum and a continuous trace of ventilatory activity may be recorded.

When ambient oxygen is very low *Arenicola* becomes quiescent, but from time to time 'tests' the water by a little pumping. Vigorous activity is resumed when oxygen in the burrow is replenished.

5.3.2   Gas transport in the blood

The circulatory system in the lugworm shows an interesting differentiation of blood flow in the branchial efferent vessels (Fig. 5–6). While all gills receive blood in the same way – partly from the ventral vessels, and partly from the body wall – blood leaves the organs of gas exchange in different directions. The posterior gills, receiving the best supply of oxygen, send blood forward through the dorsal vessel to the head and central nervous system. The anterior gills are bathed in water having a lower oxygen and higher carbon dioxide content because the ventilation stream has already given up some of its oxygen and received carbon dioxide from the posterior gills. The anterior gills send their oxygen supply to a ventral component of the gastric plexus and thence to tissues less sensitive to oxygen deprivation.

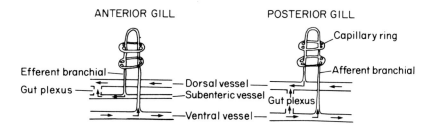

Fig. 5–6   Direction of blood flow in the vessels and gills of the lugworm, *Abarenicola affinis*. The pattern of circulation in the posterior gills takes advantage of the higher pO_2 of the incoming water to route it forward through the dorsal vessel, while the anterior gills supply tissues less sensitive to oxygen deprivation.

There is much mixing of afferent and efferent blood in the gills by direct connections through capillary rings and the principle of a countercurrent flow is not employed. Observations of live worms show each gill shrinking and swelling successively with the passage of peristaltic waves, squeezing oxygenated blood into its efferent vessels and sucking back venous blood from the afferent ones. Blood flow through the branchial system is directly dependant on the ventilatory activity of the lugworm. There is a reduction of blood flow under both the hypoxic and *hypercapnic* (when carbon dioxide tensions rise) conditions which occur during emersion.

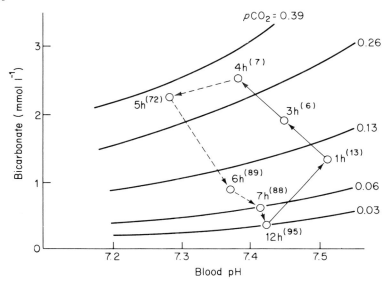

Fig. 5-7   Influence of the tides on *in vivo* values of pH and bicarbonate concentration in prebranchial blood from *Arenicola marina*. The graph shows the complex acid-base status occuring over a normal tidal cycle. The times refer to hours elapsed since the start of emersion with the corresponding measured values of haemoglobin saturation in parentheses. O——O, emersion; O– – –O, immersion. (Simplified from Toulmond, A. (1973). *Respir. Physiol.*, **19**, 130–44.)

5.3.3   Acid-base regulation

The cycle of immersion and emersion caused by the tides presents a far more complicated picture of acid-base regulation than might be seen in a mammal even when considered at constant temperature. In contrast with the crayfish, the lugworm has no simple buffer line describing the relationships between blood pH, carbon dioxide and bicarbonate over a tidal cycle (see Fig. 5–7). Important in achieving a new acid-base status with each phase of the tide is the dynamic role played by the enzyme carbonic anhydrase in transporting bicarbonate in the blood. A scheme for the removal of carbon dioxide from the tissues and its elimination across the gill surfaces is given in Fig. 5–8.

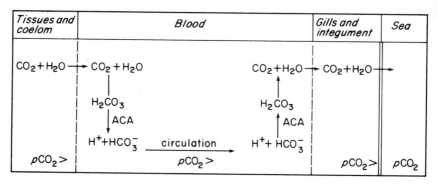

Fig. 5–8 Role of carbonic anhydrase (ACA) in the removal of metabolic carbon dioxide from the tissues of *Arenicola marina*. The enzyme reversibly catalyzes the hydration reaction of carbon dioxide depending on the partial pressure of the gas. Note that the pCO_2 gradient falls towards the external environment. (Based on Wells, R. M. G. (1973). *Comp. Biochem. Physiol.*, **46A**, 325–31.)

6 Life in Hypoxic and Anoxic Environments

The biochemical basis of respiration is covered by an earlier book in this series (BRYANT, 1980). What concerns us in this chapter are the mechanisms which enable invertebrates to sustain energy production when oxygen is limited or absent. Many invertebrates live temporarily or permanently in oxygen-free environments. Survival under these conditions may allow for the effective exploitation of food, shelter, or other resources. We have become accustomed to thinking of oxygen uptake as an exclusive reflection of an organism's energy demand. Nevertheless, whenever oxygen is in short supply and tissues become *hypoxic*, or whenever oxygen is absent and tissues are *anoxic*, there is still a continuous demand for energy. Aerobic respiration must therefore be supplemented or replaced by an alternative metabolism for which molecular oxygen is not essential. The alternative anaerobic metabolism is termed *fermentation* or *anaerobiosis*. Both aerobic and anaerobic metabolism depend on a supply of glycogen or glucose from which a high-energy phosphate compound, adenosine triphosphate (ATP) is generated. Animals accumulate glycogen, the storage form of glucose, in all tissues of the body, especially in muscles where it may be mobilized and catabolised by the process of *glycolysis* generating ATP in a sequence of reactions termed the Embden-Meyerhof-Parnas (EMP) pathway. The yield of ATP per mole of glucose catabolized is high during aerobic respiration and the end products are carbon dioxide and water. The yield is lower from anaerobiosis and the end products are organic acids. In many anaerobic tissues, the breakdown of glycogen proceeds at a faster rate than the aerobic rate. This so-called *Pasteur* effect partly compensates for the lower energy yield of anaerobic pathways and effectively conserves oxygen. For all organisms, glycolysis proceeds along a common pathway breaking down six-carbon sugars into three-carbon sugars and phosphoenolpyruvate, but beyond this intermediary compound, the end products vary in composition according to circumstances.

Until recently it was believed that the reponses of invertebrates to declining oxygen tensions necessitated a 'switch-over' from aerobic to anaerobic metabolism. Already we have drawn attention to the problem of defining a 'standard rate of metabolism', but what happens when the rate falls below this level? Actually, there is no low critical pO_2 at which 'switching' occurs and in low oxygen conditions both aerobic and anaerobic mechanisms may proceed simultaneously. The pH of invertebrate body fluids falls sharply at low oxygen tensions and this has a profound effect on the affinity and rate constants of various key enzymes, tending to favour those which participate in anaerobiosis.

One of the evolutionary determinants of the precise nature of the metabolic strategies adopted by an invertebrate is the time of exposure to anaerobic conditions. We shall now consider three strategies by which invertebrates are adapted in their own special way to oxygen deficiencies in the short, medium and long term. These animals are not eccentric offshoots from conservative aerobes, but are illustrative of the central role of the anaerobic core of metabolism that evolved long before photosynthetic organisms gave us our present atmosphere of oxygen.

6.1   Short-term anaerobiosis in insects

The essential feature of insect tracheal systems is that oxygen is diffusion limited and, under stress, delivery cannot meet total demand. Under such hypoxic conditions oxygen is rationed in favour of the central nervous system, and certain other essential tissues, at the expense of the muscles in particular, which have the capacity for supporting activity by anaerobic glycolysis. An exception is flight muscle for which the high energy demand during flight must be sustained by wholly aerobic metabolism and accordingly, this tissue is richly supplied with tracheoles.

In the common cockroach, *Periplaneta americanus*, a temporary lack of oxygen is partly compensated for by maintaining a capacity for anaerobic generation of ATP. The ATP is produced when glucose is metabolized to pyruvic and lactic acids according to the EMP pathway for glycolysis. Figure 6–1a depicts this scheme showing that there is a net gain of two moles of ATP for each mole of glucose mobilized, one-eighteenth the energy produced by respiration. This energetic strategy depends on the ultimate return of the cockroach to aerobic conditions because the accumulation of lactic acid is potentially toxic. In the presence of oxygen, lactic acid is converted back to pyruvic acid which in turn is oxidized to carbon dioxide and water with a further synthesis of ATP. If the demand for energy exceeds the oxygen supply during excessive exertion or anoxia, glycogen stores are depleted at an increased rate and lactic acid is accumulated in the tissues. The insect is then said to have incurred an oxygen debt which, when the period of oxygen deprivation is over, must be repaid by increasing the respiration rate and oxidizing the lactic acid. Decreasing the sensitivity of tissues to accumulated lactate does not appear to be a strategy employed by animals.

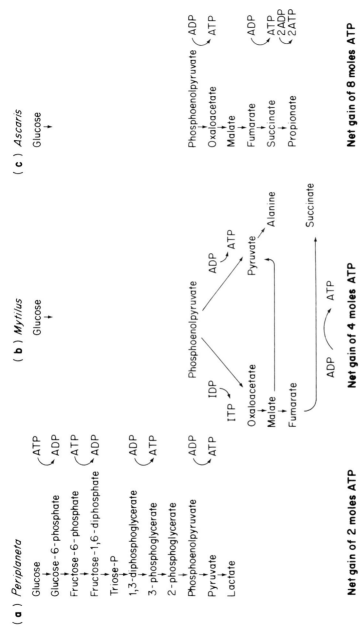

(a) *Periplaneta*

(b) *Mytilus*

(c) *Ascaris*

Net gain of 2 moles ATP

Net gain of 4 moles ATP

Net gain of 8 moles ATP

Fig. 6-1 Glycolytic strategies in three invertebrates with contrasting life-styles. Glycogen and glucose are interchangeable fuels but whereas glucose uses one ATP to enter the pathway, glycogen may be incorporated directly without the loss of an ATP. The number of moles of ATP generated from each mole of glucose is a measure of the efficiency of anaerobiosis. Since 6-carbon sugars are split into 3-carbon sugars (triose-P), the number of moles of ATP generated thereafter are doubled. Anaerobiosis in *Mytilus* involves the generation of inosine triphosphate (ITP) which is energetically equivalent to ATP. For simplicity, some intermediate compounds and the enzymes have been omitted.

The concentration of glucose in the haemolymph of insects may vary with physiological state. In *Periplaneta* and many other insects, a 'diabetogenic' factor orginating in the *corpus cardiacum* of the central nervous system acts to raise the level of blood sugar by mobilizing glycogen reserves.

Glycolysis proceeds in a similar manner in many other invertebrates and in vertebrate muscle. For this reason, the pathway leading to lactate accumulation is often termed 'classical glycolysis' in order to distinguish the pathways in other organisms that synthesize different end products.

6.2   Facultative anaerobiosis in an intertidal mollusc

It may seem odd that many intertidal invertebrates enter anaerobiosis at low tide when there is access to atmospheric oxygen. But for essentially aquatic invertebrates the desiccation stress when exposed to air may be so severe that many species isolate themselves within closed shells or burrows during ebb tides. A few bivalves such as the mussel *Mytilus edulis* may be seen to 'air gape' for brief gaseous exchange thus minimizing the period of water loss. At low tide the facultative anaerobe *Mytilus* must cease gill irrigation and remain with its valves in a tightly closed position. Obviously energy must be expended in order to maintain tension in the adductor muscle (a dead bivalve always gapes) and, in the absence of oxygen, glycolysis provides energy for this purpose. Accumulation of lactic acid might eventually impair the central nervous system and other tissues. Actually, *Mytilus* does not produce lactic acid and the less toxic end products alanine and succinate are accumulated according to a modified glycolytic scheme (Fig. 6–1b). Moreover, there is an increase in ATP generated per mole glucose catabolized when compared with the classical strategy seen in *Periplaneta*. The essential features of facultative anaerobiosis shown in the figure occur after phosphoenolpyruvate degradation.

Three processes have been favoured in the evolution of energy metabolism in *Mytilus*: firstly, enzymes such as lactate dehydrogenase have been deleted, thus avoiding the potential hazards of lactic acid accumulation; secondly, the kinetic properties of key enzymes have been modified to allow a smooth transition from respiration to anaerobiosis. Under anaerobic conditions, the tissues and body fluids of *Mytilus* become acidic because carbon dioxide cannot easily escape from the closed shell and the low pH favours the channelling of phosphoenol-pyruvate towards oxaloacetate because the enzyme pyruvate kinase which drives the reaction towards pyruvate is inhibited. Thirdly, the glycolytic reaction is coupled with other phosphorylations to increase the potential yield of high energy phosphate compounds. Despite intensive investigations into the metabolism of *Mytilus* the picture of energy production is far from complete and more complex than the scheme presented here. For example, it appears that amino acids are involved early in succinate production and that quantities of a third end product, propionate may be excreted.

One question as yet unanswered, is whether the gastrcnomic quality of a mollusc depends on its immediate metabolic history, thus does the bivalve in respiratory stress make for tough eating?

6.3   Obligatory anaerobiosis in a parasitic helminth

The intestinal roundworm, *Ascaris lumbricoides*, obtains no relief from lack of oxygen but exploits an otherwise epicurean microclimate with an abundant energy supply. Like the intertidal mollusc, *Ascaris* ferments glycogen through oxaloacetate to succinate. Beyond this, glycolysis in the parasite differs in two respects. Firstly, both pyruvate kinase and lactate dehydrogenase enzymes have been deleted because oxygen is unavailable for respiration, including repayment of oxygen debt. Secondly, because glycolysis is no longer a temporary measure, the main end product succinate is synthesized into propionate, an important product which may be incorporated in lipid metabolism. With an eight-fold increase in energy production compared with classical glycolysis, and with abundant carbohydrate supplies, *Ascaris* has an excellent adaptive strategy.

6.4   Conclusion

Anaerobic metabolism does not replace the energy yielded from respiration or maintain a high scope for activity. It does permit reduced activity with a decreased demand for energy. The length of time that an invertebrate spends under anoxic conditions is important in dictating the metabolic strategy of the species. Lactic acid is retained briefly in *Periplaneta* and an oxygen debt is incurred. A greater energy yield permits a low level of activity for several hours in *Mytilus* when the tide is out and the risks of desiccation are high. This strategy is sustained by a glycolytic pathway which leads to succinate and alanine. In the permanent absence of oxygen, the helminth *Ascaris* follows exclusively a route to succinate production which is linked into other pathways for lipid metabolism.

Further Reading

In an introductory book of this size it is not possible to give detailed references for every fact cited. Nevertheless, a number of tables and figures have been taken from the scientific literature and their sources are given in the captions.

A particularly useful source of further reading may be found in other volumes in 'Studies in Biology' and the reader is advised to scan the titles of other books in this series.

More advanced treatment of this subject is often found in large texts of Invertebrate Biology and Zoology or in Animal Physiology. However, the following references might provide additional information and at the same time, give a direct lead into the enormous volume of literature contained in scientific periodicals.

BRYANT, C. (1980), *The Biology of Respiration*, 2nd edition. Studies in Biology no. 28, Arnold, London.

DEJOURS, P. (1975). *Principles of Comparative Respiratory Physiology*. North-Holland, Amsterdam.

JONES, J. D. (1972). *Comparative Physiology of Respiration*. Arnold, London.

MANGUM, C. P. (1976). Primitive respiratory adaptations. In, *Adaptation to Environment*, R. C. Newell (ed.). Butterworths, London, pp. 191–278.

MARTIN, A. W. (1974). Circulation in invertebrates. *Annual Review of Physiology*, **36**, 171–86.

MORRIS, J. G. (1974). *A Biologist's Physical Chemistry*. Arnold, London.

NEWELL, R. C. (1973). Factors affecting the respiration of intertidal invertebrates. *American Zoologist*, **13**, 513–28.